Contents

Chapter 1: Alcohol limits and health risks

Chapter 2: Young people and alcohol

Chapter 3: Advertising

Introduction

ALCOHOL is Volume 326 in the **ISSUES** series. The aim of the series is to offer current, diverse information about important issues in our world, from a UK perspective.

ABOUT ALCOHOL

Alcohol misuse can cause many health issues. This book explores the many health problems which can arise. It looks at the risks to young people, how it can interfere with their development and the effect it can have on both their health and behaviour. It also explores the impact of living with an alcoholic parent along with how advertising can influence underage drinking.

OUR SOURCES

Titles in the **ISSUES** series are designed to function as educational resource books, providing a balanced overview of a specific subject.

The information in our books is comprised of facts, articles and opinions from many different sources, including:

⇨ Newspaper reports and opinion pieces

⇨ Website factsheets

⇨ Magazine and journal articles

⇨ Statistics and surveys

⇨ Government reports

⇨ Literature from special interest groups.

A NOTE ON CRITICAL EVALUATION

Because the information reprinted here is from a number of different sources, readers should bear in mind the origin of the text and whether the source is likely to have a particular bias when presenting information (or when conducting their research). It is hoped that, as you read about the many aspects of the issues explored in this book, you will critically evaluate the information presented.

It is important that you decide whether you are being presented with facts or opinions. Does the writer give a biased or unbiased report? If an opinion is being expressed, do you agree with the writer? Is there potential bias to the 'facts' or statistics behind an article?

ASSIGNMENTS

In the back of this book, you will find a selection of assignments designed to help you engage with the articles you have been reading and to explore your own opinions. Some tasks will take longer than others and there is a mixture of design, writing and research-based activities that you can complete alone or in a group.

Useful weblinks

www.theconversation.com

www.demos.co.uk

www.drinkaware.co.uk

www.drugs.ie

www.europa.eu

www.govuk.uk

www.the guardian.com

www.huffingtonpost.co.uk

www.ias.org.uk

www.independent.co.uk

www.kcl.ac.uk

www.kidshealth.org

www.nacoa.de

www.ncbi.nlm.gov

www.nhs.co.uk

www.nidirect.gov.uk

www.stir.ac.uk

www.thetelegraph.co.uk

FURTHER RESEARCH

At the end of each article we have listed its source and a website that you can visit if you would like to conduct your own research. Please remember to critically evaluate any sources that you consult and consider whether the information you are viewing is accurate and unbiased.

4/18

Independence Educational Publishers

First published by Independence Educational Publishers

The Studio, High Green

Great Shelford

Cambridge CB22 5EG

England

© Independence 2018

ISBN-13: 978 1 86168 777 7

Printed in Great Britain

Zenith Print Group

Alcohol "a direct cause of seven types of cancer"

"Even one glass of wine a day raises the risk of cancer: Alarming study reveals booze is linked to at least seven forms of the disease," reports MailOnline.

The news comes from a review that aimed to summarise data from a range of previous studies to evaluate the strength of evidence that alcohol causes cancer.

The main finding was that existing evidence supports the link between alcohol consumption and cancer at seven sites, including the throat, gullet, liver, colon, rectum and female breast.

The links were said to be strongest for heavy drinking, but this study suggested that even low or moderate drinking may contribute to a significant proportion of cancer cases because of how common this level of drinking is. The study also suggests there's no evidence of a "safe" level of drinking with respect to cancer.

However, it's important to be aware that this review doesn't state how the author identified and assessed the research they've drawn upon. We don't know whether all relevant research has been considered and the conclusions must be considered largely the opinion of this single author.

Nevertheless, the main finding of the link between alcohol and these seven cancers is already well recognised. Recently updated government recommendations state there's no safe level of alcohol consumption, and men and women are advised not to regularly drink more than 14 units a week. This review further supports this advice.

Where did the story come from?

The study was carried out by one researcher from the University of Otago, New Zealand. No external funding was reported.

The study was published in the peer-reviewed scientific journal *Addiction*. It is available on an open-access basis and is free to read online.

Generally the media coverage of this topic was accurate, although the tone of the reporting tended to suggest this was a new discovery, when the link between alcohol and certain types of cancer is well established.

What kind of research was this?

This was a review which aimed to summarise data from published biological and epidemiological research, and meta-analyses that have pooled data, to evaluate the strength of evidence that alcohol causes cancer.

Alcoholic drinks have been considered potentially carcinogenic (cancer causing) for a while, but there are still concerns about the validity of some observational studies finding links with cancer, and uncertainty about precisely how alcohol causes cancer.

A systematic review is the best way of gathering and summarising the available research around a particular topic area. But in this case the exact methods are not described in the paper and it's not possible to say whether they were systematic.

There's a possibility that some relevant research may have been missed and that this review is giving an incomplete picture of the issue.

What did the research involve?

The author of this review reports drawing upon biological and epidemiological research as well as meta-analyses conducted in the last ten years by a number of institutions, including the World Cancer Research Fund and American Institute for Cancer Research, the International Agency for Research on Cancer and the Global Burden of Disease Alcohol Group.

The majority of epidemiological research seemed to come from cohort and observational studies.

The research was reviewed and summarised in a narrative format which explored the evidence that alcohol causes cancer, while contrasting this with the notion that alcohol consumption may offer some form of protection from cardiovascular disease.

No methods are provided and the author does not describe how they identified the research, as you would expect from a systematic review. For example, they do not give the literature databases searched, the search dates, search terms, study inclusion or exclusion criteria, or descriptions of how studies were quality assessed.

What were the basic results?

There were several findings from this study, the main one being that existing evidence supports the link between alcohol consumption and cancer at seven sites: oropharynx (mouth and throat), larynx (voice box), oesophagus (gullet), liver, colon (bowel), rectum and female breast.

The strength of the association differed by the site of the cancer. It was strongest for the mouth, throat and oesophagus, with the review suggesting that someone who drinks more than 50g of alcohol a day is four to seven times more likely to develop these types of cancer compared to someone who doesn't drink. As the author says, the interaction of smoking with alcohol is also believed to contribute to the risk of these cancers.

The link was comparatively weaker for colorectal, liver and breast cancer. The review suggests someone who drinks more than 50g of alcohol a day is 1.5 times more likely to develop these types of cancer compared to someone who doesn't drink.

For all of these associations there was a dose-response relationship, where increased consumption was linked with an increase in cancer risk. This applied to all types of alcoholic drinks. The highest risks were associated with heavier drinking. There was also some suggestion that the level of risk goes down over time when alcohol consumption stops.

Recent large studies have found uncertain evidence whether low to moderate consumption has a significant effect on total cancer risk. But given that this level of consumption is common in the general population, the author considers that it could still contribute to a significant number of cases.

Furthermore, they say there is no clear threshold of what constitutes a harmful level of alcohol consumption, and therefore no safe level of drinking with respect to cancer.

The author also suggests that confounding factors may be responsible for the protective effect between alcohol consumption and cardiovascular disease that has been found in previous studies. For example, this may be due to the potential bias caused by misclassification of former drinkers as abstainers.

The research went on to report that alcohol is estimated to be responsible for approximately half a million deaths from cancer in 2012 and 5.8% of cancer deaths worldwide, deeming it to be a significant public health burden.

How did the researchers interpret the results?

The author concluded: "There is strong evidence that alcohol causes cancer at seven sites, and probably others. The measured associations exhibit gradients of effect that are biologically plausible, and there is some evidence of reversibility of risk in laryngeal, pharyngeal and liver cancers when consumption ceases."

"The highest risks are associated with the heaviest drinking, but a considerable burden is experienced by drinkers with low to moderate consumption, due to the distribution of drinking in the population."

Conclusion

This narrative review aimed to summarise data from published biological and epidemiological research to discuss the strength of evidence that alcohol causes cancer.

The author gives their main finding as a link between alcohol consumption and cancer at seven sites, and also that the highest risks seem to be associated with heavier drinking. However, they state there's no "safe" drinking threshold and that low to moderate consumption still contributes to a significant number of cancer cases.

The biggest limitation of this review is that it doesn't appear to be systematic. The author provided no methods for how they identified and appraised the research they drew on. Despite referencing a number of large studies and reviews, this study and its conclusions have to be considered largely the opinion of the author following their appraisal of the evidence.

We don't know whether the review has considered all research relevant to the topic and is able to reliably quantify the risks of cancer – overall or at specific sites – associated with alcohol consumption.

An additional limitation to keep in mind is that this data mainly appeared to be from observational studies. These cannot prove cause and effect. The individual studies will likely have varied considerably in the additional health and lifestyle factors they took account of when looking at the links with alcohol. For example, smoking, diet and physical activity are all factors likely to be associated both with level of alcohol consumption and cancer risk.

As the author notes in particular, confounding factors may be responsible for the observed protective effect between alcohol consumption and cardiovascular disease.

Another limitation is that alcohol consumption is likely to be self-reported in the studies analysed, which may be inaccurate and lead to misclassification. For example, a potential bias that the author notes is classifying former drinkers as abstainers.

The author does consider the limitations of these observational findings, saying: "The limitations of cohort studies mean that the true effects may be somewhat weaker or stronger than estimated currently, but are unlikely to be qualitatively different."

But despite the methodological limitations of this review, it does support current understanding around this topic. Cancer Research UK also reports that alcohol can increase risk of these seven cancers and that there is no "safe" alcohol limit.

While we can't give a safe limit to drink when it comes to cancer, people are advised to follow current alcohol recommendations, which are to drink no more than 14 units per week and to spread your drinking over three days or more if you drink as much as 14 units a week.

22 July 2016

⇨ The above information is reprinted with kind permission from NHS Choices. Please visit www.nhs.co.uk for further information.

New alcohol guidelines show increased risk of cancer

Updated alcohol consumption guidelines give new advice on limits for men and pregnant women.

New guidelines for alcohol consumption, produced by the UK Chief Medical Officers, warn that drinking any level of alcohol increases the risk of a range of cancers. This is supported by a new review from the Committee on Carcinogenicity (CoC) on alcohol and cancer risk.

It is now known that the risks start from any level of regular drinking and increase with the amount being drunk, and the new guidelines are aimed at keeping the risk of mortality from cancers or other diseases low. The links between alcohol and cancer were not fully understood in the original guidelines, which came out in 1995.

This review also found that the benefits of alcohol for heart health only apply for women aged 55 and over. The greatest benefit is seen when these women limit their intake to around five units a week, the equivalent of around two standard glasses of wine. The group concluded that there is no justification for drinking for health reasons.

These issues prompted changes to alcohol guidelines for men. Men should not drink more than 14 units of alcohol each week, the same level as for women. This equals six pints of average strength beer a week, which would mean a low risk of illnesses such as liver disease or cancer. The previous guidelines were 21 units for men and 14 units for women per week.

An additional recommendation is not to 'save up' the 14 units for one or two days, but to spread them over three or more days. People who have one or two heavy drinking sessions each week increase the risk of death from long-term illnesses, accidents and injuries. A good way to reduce alcohol intake is to have several alcohol-free days a week.

The guidelines for pregnant women have also been updated to clarify that no level of alcohol is safe to drink in pregnancy. The previous advice for pregnant women to limit themselves to no more than one to two units of alcohol once or twice per week has been removed to provide greater clarity as a precaution.

Dame Sally Davies, Chief Medical Officer for England, said: "Drinking any level of alcohol regularly carries a health risk for anyone, but if men and women limit their intake to no more than 14 units a week it keeps the risk of illness like cancer and liver disease low.

"I want pregnant women to be very clear that they should avoid alcohol as a precaution. Although the risk of harm to the baby is low if they have drunk small amounts of alcohol before becoming aware of the pregnancy, there is no 'safe' level of alcohol to drink when you are pregnant."

What we are aiming to do with these guidelines is give the public the latest and most up to date scientific information so that they can make informed decisions about their own drinking and the level of risk they are prepared to take.

Advice on single episodes of drinking is included in the guidelines for the first time. To keep the short-term health risks low:

⇨ limit the total amount of alcohol drunk on any one occasion

⇨ drink more slowly, with food and alternate with water.

Guidance about a set number of units for a single occasion or day was considered. However, partly because the short-term risks for individuals vary so widely, this is not currently included in the guidelines. Whether the new guidelines should include a specific number of units, as a simple number can be easier to follow than more general advice, is included in the consultation.

This new advice follows a detailed review of the scientific evidence used for the guidelines in 1995. This work has been underway since 2013, led by a panel of experts in public health, behavioural science and alcohol studies.

The development of the new guidelines was chaired by Professor Mark Petticrew, Professor of Public Health Evaluation at the London School of Hygiene and Tropical Medicine and by Professor Sally Macintyre, Professor Emeritus at the University of Glasgow.

Professor Petticrew, said:

"This new guidance has been based on a wide range of new evidence from this country and overseas. We have reviewed all the evidence thoroughly and our guidance is firmly based on the science, but we also considered what is likely to be acceptable as a low risk level of drinking and the need to have a clear message."

The CoC's latest findings support the significant links from alcohol to cancer outlined in the new guidance.

The CoC review found that drinking even a small amount of alcohol increases the risk of some cancers compared with people who do not drink at all. The risk of getting some alcohol-related cancers gradually reduces over time when people stop drinking, but can take many years before the risk falls to the levels found in people who have never drunk alcohol.

8 January 2016

⇨ The above information is reprinted with kind permission from GOV. UK. Please visit www.govuk.uk for further information.

Health check: does drinking alcohol kill the germs it comes into contact with?

***An article from* The Conversation.**

THE CONVERSATION

By Vincent Ho, Lecturer and Clinical Academic Gastroenterologist, Western Sydney University

Alcohol is a well-known disinfectant and some have speculated it may be useful for treating gut infections. Could alcohol be a useful agent to treat tummy bugs and throat infections?

Wine has long been known for its disinfecting and cleansing properties. According to historical records, in the third century AD Roman generals recommended wine to their soldiers to help prevent dysentery.

Can alcohol kill germs in our guts and mouths?

Wine was examined as part of a 1988 study that tested a number of common beverages (carbonated drinks, wine, beer, skim milk and water) for their antibacterial effect. The beverages were inoculated with infectious gut bacteria such as salmonella, shigella and E. coli. After two days it was found the organisms fared worst in red wine.

Beer and carbonated drinks had an effect but were not as effective as wine.

A number of years later a laboratory study was carried out to work out what in wine was causing the antibacterial effect. The researchers tested red wine on salmonella and compared it to a solution containing the same alcohol concentration and pH level (acidic).

Red wine was seen to possess intense antibacterial activity, which was greater than the solution with the same concentration of alcohol and pH. Even though a large proportion of the antibacterial effect of red wine against salmonella was found to be due to its acid pH and alcohol concentration, these factors only partly explained the observed effects.

The concentration of alcohol is certainly important for the effect on bugs (microbes). For alcohol hand rubs a high alcohol concentration in the range of 60–80% is considered optimal for antimicrobial activity.

A laboratory study looked at the penetration of alcohol into groups of microorganisms in the mouth and its effect on killing microbes. Alcohol concentrations lower than 40% were found to be significantly weaker in affecting bacterial growth. Alcohol with a 10% concentration had almost no effect.

The exposure time of alcohol was also important. When 40% alcohol (the same concentration as vodka) was used the effect on inhibiting the growth of these microorganisms was much greater when applied over 15 minutes compared to six minutes. It was determined that 40% alcohol had some ability to kill oral bacteria with an exposure time of at least one minute.

Can alcohol damage the stomach?

In a study involving 47 healthy human volunteers, different alcohol

concentrations (4%, 10%, 40%) or saline, as a control, were directly sprayed on the lower part of the stomach during a gastroscopy (where a camera is inserted down into the stomach through the mouth).

The greater the concentration of alcohol, the more damage was observed in the stomach. Erosions accompanied by blood were the typical damage observed in the stomach. No damage was observed in the small bowel. Stomach injury caused by higher alcohol concentrations (greater than 10%) took more than 24 hours to heal.

So in theory a high enough concentration of alcohol swallowed (or kept in the mouth for at least a minute) would kill a large number of gut and oral bacteria, but it would very likely do some damage to the stomach lining.

Chronic use of alcohol can also lead to an overgrowth of bacteria in the small bowel. This has been thought to be linked to gastrointestinal symptoms such as diarrhoea, nausea and vomiting, which are frequently noted in alcoholic patients.

So what's the verdict?

Alcohol consumption can lead to some immediate damage to the gut, with greater damage seen at higher concentrations. In theory a high enough alcohol concentration with sufficient exposure to gut or oral tissue could kill bacteria but will in all likelihood also damage the gut lining.

It's not advised alcohol be used as a regular disinfectant to treat tummy bugs or throat infections.

11 October 2017

⇨ The above information is reprinted with kind permission from *The Conversation*. Please visit www.theconversation.com for further information.

Alcoholism could be in your genes, not your brain

But there is no single 'alcoholism' gene.

By Sophie Gallagher

New evidence furthers the claim that alcoholism is determined by your genes, and increases the likelihood that in the future we will be able to identify potential candidates likely to suffer.

Researchers at Indiana University School of Medicine have identified hundreds of genes and genetic codes in rats that appear to play a role in increasing the desire to drink alcohol.

In the trials rats were carefully bred to either drink large amounts of alcohol or to reject it (mimicking all criteria of human alcoholism).

By breeding these animals specifically for these experiments, scientists were able to avoid issues that have plagued other alcoholism studies.

These issues included the inability to account for all family history of drinking and non-genetic economic, social and cultural factors that lead to heavy drinking.

After the rats had been bred, they were subject to genome analysis that found the drinking and non-drinking groups of rats had key regions of genetic code that differed from each other.

These were large segments of code, rather than any particular one 'alcoholism' gene.

Lead researcher William Muir, said: "This research highlights that alcoholism in rats has a strong genetic component and is influenced by many hundreds of genes, each with small effects.

"There is no single gene responsible for alcoholism. However, critical regulatory pathways involving several of the genes discovered were found."

These findings, published in the *PLOS Genetics* journal, suggest that there are potential pharmalogical solutions to alcoholism in humans in the future.

And medical professionals will be able to accurately predict those at high risk of becoming alcoholics.

5 August 2016

⇨ The above information is reprinted with kind permission from *Huffington Post*. Please visit www.huffingtonpost.co.uk for further information.

The risks of drinking too much

Regularly drinking more than 14 units a week risks damaging your health.

14 units is equivalent to six pints of average-strength beer or ten small glasses of low-strength wine.

New evidence around the health harms from regular drinking have emerged in recent years.

There is now a better understanding of the link between drinking and some illnesses, including a range of cancers.

The previously held position that some level of alcohol was good for the heart has been revised.

It is now thought that the evidence on a protective effect from moderate drinking is less strong than previously thought.

Low-risk drinking advice

To keep health risks from alcohol to a low level if you drink most weeks:

⇨ men and women are advised not to drink more than 14 units a week on a regular basis

⇨ spread your drinking over three or more days if you regularly drink as much as 14 units a week

⇨ if you want to cut down, try to have several drink-free days each week.

If you're pregnant or think you could become pregnant, the safest approach is not to drink alcohol at all to keep risks to your baby to a minimum.

No 'safe' drinking level

If you drink less than 14 units a week, this is considered low-risk drinking.

It's called "low risk" rather than "safe" because there is no safe drinking level.

The type of illnesses you can develop after ten to 20 years of regularly drinking more than 14 units a week include:

⇨ cancers of the mouth, throat and breast

⇨ stroke

⇨ heart disease

⇨ liver disease

⇨ brain damage

⇨ damage to the nervous system.

The effects of alcohol on your health will depend on how much you drink. The less you drink, the lower the health risks.

Read about alcohol units to work out how much alcohol there is in your drinks.

'Single session' drinking

Drinking too much too quickly on any single occasion can increase your risk of:

⇨ accidents resulting in injury, causing death in some cases

⇨ misjudging risky situations

⇨ losing self-control, like having unprotected sex.

To reduce your health risks on any single session:

⇨ limit how much you drink

⇨ drink more slowly

⇨ drink with food

⇨ alternate with water or non-alcoholic drinks.

⇨ The above information if reprinted with kind permission from NHS Choices. Please visit www.nhs.uk for further information.

© NHS Choices 2018

The Guardian view on drinking culture: statistics, myths and alcohol abuse

The last Friday before Christmas is, wait for it, Booze Black Friday. Not, for once, a new marketing tool, but the gloomy prognosis of the country's ambulance services as they await the predictable consequences of the day when it's estimated that alcohol sales peak and Britons double their normal alcohol consumption. All over the Christmas-celebrating world, but particularly in the countries of northern Europe and in Australia, there will be unhappy people waking up tomorrow feeling, as the hero of Kingsley Amis's *Lucky Jim* felt, as if "his mouth had been used as a latrine by some small creature of the night, and then as its mausoleum".

Drinking so much alcohol that it becomes a health concern is largely restricted to the parts of the globe where alcohol is available, acceptable and affordable, and good health the norm. In these countries, regular harmful drinking (anything above the recommended daily units of alcohol – three for women and four for men) is most common among middle-class, middle-aged people, particularly men but increasingly women, and among unemployed men. Binge drinkers tend to be young. The growth in excessive drinking that has been evident for a generation may have peaked; all the same, according to government figures drinking too much costs the NHS £3.5 billion, productivity falls by £7.5 billion and drink-associated crime costs a further £11 billion. However bad the hangover ("even my hair hurts": Rock Hudson, *Pillow Talk*), drinking is still often seen as glamorous, and it still predisposes drinkers to very unglamorous diseases such as cirrhosis of the liver, various cancers and a tendency to pick fights. Efforts are struggling to persuade tipplers that they are not, as Samuel Johnson said, improved, merely unaware of their defects. No wonder campaigners want minimum pricing of alcohol, something the Government rejects despite its own research showing a link between higher prices and less harmful drinking. Just this week, the *BMJ* reported that the advisory body NICE had been warned by official sources not to discuss alcohol pricing.

The evidence linking consumption and harm is unquestionable. All the same, in the UK, there is a whiff of moral panic about the reporting of the way people drink. Take the claim that Britons who drink, drink more than in the rest of the drinking world. Not so. Germany and France, as well as most of eastern Europe (and don't even mention Russia) all have higher average consumption. The Organisation for Economic Cooperation and Development average is 10.1 litres of alcohol per capita. While in the US, most people drink much less (and many people don't drink at all), in the UK, the average is 10.6 litres, but in France it is a whole litre more. In all countries, the tendency is for the top one-fifth of drinkers to consume more than half of all the alcohol that's drunk.

So while it's unquestionably the case that some people drink far too much for their own or society's good and ought to rein in or stop, most people don't overdo it. For every problem drinker, four people take only a glass or two from time to time. What marks Britons out is the widespread determination to go out and get smashed. It should not be a badge of honour to end a night out in the booze tent put up at the station by a desperate NHS ambulance trust. Remember Samuel Butler's observation: "If the headache would only precede the intoxication, alcoholism would be a virtue." It doesn't, and it isn't.

18 December 2015

⇨ The above information is reprinted with kind permission from *The Guardian*. Please visit www.theguardian.com for further information.

How safe alcohol guidelines vary around the world

Bottoms up.

By Rachel Moss

If you're looking to monitor your alcohol intake, you may be likely to look up the Government's recommendations around safe consumption.

But a new review has revealed that official guidelines around low-risk drinking vary greatly around the world.

Researchers from Stanford University found that measurements of the amount of alcohol in a "standard drink" ranged from 8g in Iceland and the UK to 20g in Austria.

In the most conservative countries, "low-risk" consumption means drinking no more than 10g of alcohol per day for women and 20g for men.

But in Chile, a person can down 56g of alcohol per day and still be considered a low-risk drinker.

The scientists analysed the definitions of "standard drink" and "low-risk" drinking in 37 countries around the world to create their report.

They concluded that there is a risk of confusion around safe drinking due to the varying figures.

Psychiatrist Professor Keith Humphreys, who co-led the research, told the Press Association: "There's a substantial chance for misunderstanding.

"A study of the health effects of low-risk drinking in France could be misinterpreted by researchers in the United States who may use a different definition of drinking levels.

"Inconsistent guidelines are also likely to increase scepticism among the public about their accuracy. It is not possible that every country is correct; maybe they are all wrong."

He added: "If you think your country should have a different definition of a standard drink or low-risk drinking, take heart – there's probably another country that agrees with you."

The number of units you are drinking depends on the size and strength of your drink

11% ABV wine	14% ABV wine	2.8% ABV lager	4.8% ABV lager
1.4 units	1.8 units	0.8 units	1.4 units
1.9 units	2.5 units	1.2 units	2.1 units
2.8 units	3.5 units	1.6 units	2.7 units
8.3 units	10.5 units	1.8 units	3.2 units

Source: gov.scot

In the UK, new advice introduced in January says men should not drink more than 14 units of alcohol per week, the same as the limit for women. The previous guidelines were 21 units for men and 14 units for women per week.

Adding to confusion is the difference between a unit and a gram of alcohol.

A unit translates to 10ml, or 8g, of pure alcohol – the amount of alcohol the average adult can process in an hour.

A one-unit alcoholic drink is roughly equivalent to 250ml of 4% strength beer, 76ml of 13% wine or 25ml of 40% spirits.

According to the latest report, although the World Health Organization (WHO) defines a standard drink as one containing 10g of alcohol, this is not accepted by half the countries studied.

Professor Humphreys said: "More and more countries are trying to give their citizens guidelines about how much alcohol is safe to drink, and for whom.

At the very least, we should know whether it's true that women should drink less than men. But even this is unclear.

"We've also learned that what constitutes a 'standard drink' in each country is far from standard, despite the WHO's recommendation.

"But in many cases these guidelines are adopted as public health policy and even printed onto alcoholic beverages without knowing whether people read them, understand them or change their behaviour as a result."

The study is published in full in the journal *Addiction*.

13 April 2016

⇨ The above information is reprinted with kind permission from *Huffington Post*. Please visit www.huffingtonpost.co.uk for further information.

Fewer adults dying from conditions directly caused by alcohol, but deaths from related conditions, such as cancer, rise by 1%

New figures from Public Health England show fewer adults are dying from alcohol-specific conditions, such as alcoholic liver disease and alcohol poisonings.

The latest update to the Local Alcohol Profiles for England (LAPE) data tool shows that nationally, alcohol-specific deaths fell by 3% to 17,755 deaths. Alcohol-related deaths have seen a slight increase, year on year, from an estimated 22,330 in 2012 to 22,976 in 2014. Alcohol-related deaths include conditions that are partially related to alcohol, such as heart disease and certain cancers.

A 3% decrease in alcohol-specific deaths is promising; however, a lot of the ill health we are seeing associated with alcohol, such as heart disease and cancer, is among people who are not dependent, but who drink frequently and are unaware of the risks. In both alcohol-specific and alcohol-related death rates, the rate for men is almost double that of women.

For the first time, the LAPE tool includes data on alcohol-related road traffic accidents. This shows that between 2012 and 2014, in 2.6% of reported road traffic accidents, one or more driver failed a breath test.

There continues to be large variations in alcohol-related harms across the country, with 165 local authorities seeing an increase in alcohol-related deaths in 2014 and 161 seeing a drop.

Substantial health inequalities continue to exist for both men and women, with the rate of liver disease in the most deprived areas double the rate in the least deprived.

Professor Kevin Fenton, Director of Health and Wellbeing at PHE said: ""There are over ten million people in England drinking alcohol at increasingly harmful levels putting them at risk of conditions such as cancer. For women who drink, they are 20% more likely to get breast cancer than those that don't."

Alcohol harms individuals, families and communities and it's crucial that, alongside effective local interventions and treatment for those that need it, we look more widely at what affects drinking behaviour in this country. Public Health England will soon be providing a report to the Government on how we can reduce the harms caused by alcohol.

The LAPE tool presents data for 23 alcohol-related indicators in an interactive tool, which helps local areas assess alcohol-related harm and monitor the progress of efforts to reduce this.

Background information

The Local Alcohol Profiles for England have been updated with 2014 data, including by local authority area and region. Main findings include:

⇨ alcohol-specific deaths fell to 17,755 in the three-year period 2012 to 2014. Down 3% compared to the previous three-year period (3.1% fall among men and a 2.5% fall among women)

⇨ alcohol-related mortality rose slightly (by 0.8%) to 22,967 deaths in 2014, compared to 22,779 in 2013

⇨ in both alcohol-specific and alcohol-related mortality, the rate for men is almost double that of women

⇨ the mortality rate from chronic liver disease remains unchanged (17,238 deaths in 2012 to 2014) but has reduced since 2006, showing signs of a downward trend (a 7% drop overall since the start of the LAPE series in 2006 to 2008) which would bring England in line with other European countries

⇨ the rate of alcohol-related road traffic accidents in England fell by 5% (to 26.4 per 1,000 road traffic accidents, 10,157 accidents) for the latest time period (2012 to 2014) compared to the previous period (2011 to 2013)

The Chief Medical Officer alcohol guidelines have been updated and are under consultation, advising men and women to drink no more than 14 units of alcohol per week.

Public Health England exists to protect and improve the nation's health and well-being, and reduce health inequalities. It does this through world-class science, knowledge and intelligence, advocacy, partnerships and the delivery of specialist public health services. PHE is an operationally autonomous executive agency of the Department of Health.

11 March 2016

⇨ The above information is reprinted with kind permission from GOV. UK. Please visit www.gov.uk for further information.

© Crown copyright 2018

New alcohol advice issued

New proposed guidelines on alcohol, drawn up by the Chief Medical Officers of the UK, have been published today.

The expert group that produced the guidelines looked at the body of new evidence about the potential harms of alcohol that has emerged since the previous guidelines were published in 1995.

There are three main issues on which revised or new guidance is given:

⇨ guidance on regular drinking

⇨ guidance on single drinking sessions

⇨ guidance on drinking in pregnancy.

Regular drinking

The guidance advises that:

⇨ to keep health risks from drinking alcohol to a low level you are safest not regularly drinking more than 14 units per week – 14 units is equivalent to a bottle and a half of wine or five pints of export-type lager (5% abv) over the course of a week – this applies to both men and women

⇨ if you do drink as much as 14 units per week, it is best to spread this evenly over three days or more

⇨ if you have one or two heavy drinking sessions, you increase your risks of death from long-term illnesses and from accidents and injuries

⇨ the risk of developing a range of illnesses (including, for example, cancers of the mouth, throat and breast) increases with any amount you drink on a regular basis

⇨ if you wish to cut down the amount you're drinking, a good way to achieve this is to have several alcohol-free days each week.

Single drinking sessions

The new proposed guidelines also look at the potential risks of single drinking sessions, which can include accidents resulting in injury (causing death in some cases), misjudging risky situations, and losing self-control.

You can reduce these risks by:

⇨ limiting the total amount of alcohol you drink on any occasion

⇨ drinking more slowly, drinking with food, and alternating alcoholic drinks with water

⇨ avoiding risky places and activities, making sure you have people you know around, and ensuring you can get home safely.

Some groups of people are more likely to be affected by alcohol and should be more careful of their level of drinking. These include:

⇨ young adults

⇨ older people

⇨ those with low body weight

⇨ those with other health problems

⇨ those on medicines or other drugs.

Drinking and pregnancy

The guidelines recommend that:

⇨ if you are pregnant or planning a pregnancy, the safest approach is not to drink alcohol at all, to keep risks to your baby to a minimum

⇨ drinking in pregnancy can lead to long-term harm to the baby, with the more you drink the greater the risk

If you have just discovered you are pregnant and you have been drinking then you shouldn't automatically panic as it is unlikely in most cases that your baby has been affected; though it is important to avoid further drinking.

If you are worried about how much you have been drinking when pregnant, talk to your doctor or midwife.

Why have the guidelines been revised?

There are a number of factors that have come to light since 1995 or were thought important by the expert group so they needed to be highlighted to the public. These include:

⇨ The benefits of moderate drinking for heart health are not as strong as previously thought and apply to a smaller proportion of the population – specifically women over the age of 55. In addition there are more effective methods of increasing your heart health, such as exercise.

⇨ The risks of cancers associated with drinking alcohol were not fully understood in 1995. Taking these risks on board, we can no longer say that there is such a thing as a "safe" level of drinking. There is only a "low risk" level of drinking.

⇨ The previous guidelines did not address the short-term risks of drinking, especially heavy drinking, such as accidental head injury and fractures.

⇨ In pregnancy the expert group thought a precautionary approach was best and it should be made clear to the public that it is safest to avoid drinking in pregnancy.

Dame Sally Davies, Chief Medical Officer for England, said: "What we are aiming to do with these guidelines is give the public the latest and most up-to-date scientific information so that they can make informed decisions about their own drinking and the level of risk they are prepared to take."

The proposed guidance comes into effect from 8 January. The consultation is due to finish by 1 April 2016 and will seek the public's view on how helpful and easy to use the new advice is, not the scientific basis for it.

Read more about the risks of drinking too much and advice on how to cut down on your drinking.

8 January 2016

⇨ The above information is reprinted with kind permission from NHS Choices. Please visit www.nhs.uk for further information.

© NHS Choices 2018

Chapter 2

Young people and alcohol – what are the risks?

The effects of alcohol on young people are not the same as they are on adults. While alcohol misuse can present health risks and cause careless behaviour in all age groups, it is even more dangerous for young people. Find out how alcohol can affect young people's health and behaviour.

Health risks

Because young people's bodies are still growing, alcohol can interfere with their development. This makes young people particularly vulnerable to the long-term damage caused by alcohol. This damage can include:

⇨ cancer of the mouth and throat

⇨ sexual and mental health problems, including depression and suicidal thoughts

⇨ liver cirrhosis and heart disease.

Research also suggests that drinking alcohol in adolescence can harm the development of the brain.

Young people might think that any damage to their health caused by drinking lies so far in the future that it's not worth worrying about. However, there has been a sharp increase in the number of people in their twenties dying from liver disease as a result of drinking heavily in their teens.

Young people who drink are also much more likely to be involved in an accident and end up in hospital.

Risky behaviour – sex

Drinking alcohol lowers people's inhibitions, and makes them more likely to do things that they would not normally do. Young people are particularly at risk because at their stage of life, they are still testing the boundaries of what is acceptable behaviour.

One in five girls (and one in ten boys) aged 14 to 15 goes further than they wanted to in a sexual experience after drinking alcohol. In the most serious cases, alcohol could lead to them becoming the victim of a sexual assault.

Unsafe sex and unwanted pregnancy

If young people drink alcohol, they are more likely to be reckless and not use contraception if they have sex. Almost one in ten boys and around one in eight girls aged 15 to 16 have unsafe sex after drinking alcohol. This puts them at risk of sexual infections and unwanted pregnancy.

Research shows that a girl who drinks alcohol is more than twice as likely to have an unwanted pregnancy as a girl who doesn't drink.

Antisocial behaviour

Alcohol interferes with the way people think and makes them far more likely to act carelessly. If young people drink alcohol, they are more likely to end up in dangerous situations.

For example, they are more likely to climb walls or other heights and fall off. Or they might verbally abuse someone who could hit them. They are also more likely to become aggressive themselves and throw a punch.

Four out of ten secondary school-age children have been involved in some form of violence because of alcohol. This could mean they have been beaten up or robbed after they've been drinking, or have assaulted someone themselves.

Getting into trouble with the police

If a child or young person drinks alcohol, then they are more likely to get into trouble with the police. Every year in the UK, more than 10,000 fines for being drunk and disorderly are issued to young people aged 16 to 19.

Children as young as 12 are being charged with criminal damage to other people's property as a result of drinking.

Criminal behaviour

Young people who get drunk at least once a month are twice as likely to commit a criminal offence as those who don't. More than one in three teenagers who drink alcohol at least once a week have committed violent offences such as robbery or assault.

Young people who get involved with crime are also likely to end up with a criminal record. This can damage their prospects for the rest of their life. Having a criminal record can prevent people from getting some jobs and, for some offences, prevent them from travelling abroad.

Failing to meet potential at school

When young people drink, it takes longer for the alcohol to get out of their system than it does in adults. So if young people drink alcohol on a night before school, then they can do less well in lessons the next day.

Young people who regularly drink alcohol are twice as likely to miss school and get poor grades as those who don't. Almost half of young people excluded from school in the UK are regular drinkers.

⇨ The above information is reprinted with kind permission from Nidirect. Please visit www.nidirect.gov.uk for further information.

Should my child drink alcohol?

Children and young people are advised not to drink alcohol before the age of 18.

Alcohol use during the teenage years is related to a wide range of health and social problems.

However, if children do drink alcohol underage, it shouldn't be until they are at least 15.

Health advice

The Chief Medical Officer has provided guidance on the consumption of alcohol by children and young people. This can help parents make decisions about their children and their relationship with alcohol.

Health risks:

⇨ Drinking alcohol can damage a child's health, even if they're 15 or older. It can affect the normal development of vital organs and functions, including the brain, liver, bones and hormones.

⇨ Beginning to drink before age 14 is associated with increased health risks, including alcohol-related injuries, involvement in violence, and suicidal thoughts and attempts.

⇨ Drinking at an early age is also associated with risky behaviour, such as violence, having more sexual partners, pregnancy, using drugs, employment problems and drink driving.

Advice for parents

⇨ If children do drink alcohol, they shouldn't do so until they're at least 15 years old.

⇨ If 15–17 year-olds drink alcohol, it should be rarely and never more than once a week. They should always be supervised by a parent or carer.

⇨ If 15–17 year olds drink alcohol, they should never exceed the recommended adult weekly limit (14 units of alcohol). One unit of alcohol is about half a pint of normal-strength beer or a single measure (25ml) of spirits. A small glass of wine equals 1.5 units of alcohol. Read more about alcohol units.

⇨ If your child intends to drink alcohol, using positive practices such as incentives, setting limits, agreeing on specific boundaries and offering advice can help.

Talking to your child

Talk to your child about the dangers of alcohol before they start drinking. You can use the points below as guidance.

⇨ Make it clear that you disapprove. Research suggests that children are less likely to drink alcohol when their parents show that they don't agree with it.

⇨ Don't shout at your child, because it will make them defensive and could make the situation worse. Stay calm and firm.

⇨ Make it clear that you're there for them if they need you, and answer any questions they have.

⇨ Talk to your child about how alcohol affects judgement. Drinking too much could lead them to doing something they later regret, such as having unprotected sex, getting into fights or drink driving.

⇨ Warn your child about the dangers of drink spiking and how to avoid it.

⇨ If your child wants to drink alcohol, advise them to eat something first, not drink too much and have a soft drink between alcoholic drinks.

⇨ Make sure your child tells you where they're going and has a plan for getting home safely. If they're planning to drink, make sure they're with friends who can look after them.

Drinkaware also has information and advice about talking to your child about alcohol.

What the law says

The police can stop, fine or arrest a person under 18 who is drinking alcohol in public. If you're under 18, it's against the law:

⇨ for someone to sell you alcohol

⇨ to buy or try to buy alcohol

⇨ for an adult to buy or try to buy alcohol for you

⇨ to drink alcohol in licensed premises, such as a pub or restaurant.

However, if you're 16 or 17 and accompanied by an adult, you can drink (but not buy) beer, wine or cider with a meal.

If you're 16 or under, you may be able to go to a pub or premises that's primarily used to sell alcohol if you're accompanied by an adult. However, this isn't always the case and it can depend on the premises and the licensable activities taking place there.

It's illegal to give alcohol to children under five.

⇨ The above information is reprinted with kind permission from NHS Choices. Please visit www.nhs.uk for further information.

Alcohol Awareness Week: how to talk to your kids about alcohol

Do you discuss your hangovers with your kids?

When you think about the important life lessons you need to teach your kids as they grow up, a conversation about alcohol probably isn't the top of that list.

But as it is unlikely children will make it to the legal drinking age before they start to form their own opinions on alcohol, experts in the field believe it is crucial parents have opened up this topic of conversation at an early age.

The focus for this year's Alcohol Awareness Week [13–19 November] is on families, specifically addressing the stigma that can keep teens from hiding the truth about their drinking habits, by having open and honest conversations about drinking.

Vivienne Evans, chief executive of the families, drugs and alcohol charity Adfam told HuffPost UK: "Alcohol is part of our culture in this country and is something children and young people are exposed to through advertising, their peers and their parents."

Evans continued: "Parents sometimes don't realise how much their children look up to them and how much they shape later behaviour as adults.

"We think it's therefore important that all parents talk to their children openly and honestly to equip them with the knowledge and skills they need to make safe decisions."

The Institute of Alcohol Studies' (IAS) 2017 report states that English Chief Medical Officers recommend a child does not drink before the age 15, and between the ages of 15 and 18 drinking should be supervised by an adult.

However, the the latest figures from NHS show 44% of 11- to 15-year-olds have tried alcohol. So it's evident this is a conversation that needs to be had at an early age.

Here are six guidelines parents should follow when starting that conversation, according to experts at alcohol organisations Drinkaware and Adfam.

1. Get the timing right.

Drinkaware's advice is that the earlier you bring up the topic with your children, the better. But remember: timing is everything.

"Starting a discussion just as your child is going out the door to meet friends, before bed, or in the middle of an argument about other things can lead to conflict," said Dr Sarah Jarvis, medical advisor to Drinkaware.

2. Make it an ongoing conversation.

Just like the 'sex' talk, the 'alcohol' talk shouldn't be a one-off conversation with your children, but rather one that is frequently addressed.

"You're more likely to have a greater impact on your child's decisions about drinking if you have a number of chats as part of an ongoing conversation," said Dr Jarvis.

This includes bringing up the topic of alcohol in relation to topical events, as Dr Jarvis added: "If your children haven't brought up the subject you could find a 'hook' – a recent film or TV storyline, a celebrity scandal involving drink, even stories about family or friends – simply ask: 'What do you think?' and follow on from what they say."

3. Be honest about your own alcohol consumption.

Dr Jarvis said it is likely your child may ask about your own alcohol consumption. Rather than brush it off, help them understand your drinking habits, as well as addressing the risks. Do so in an educational manner.

"Children aren't stupid," Dr Jarvis told HuffPost UK. "If you claim you never drink or never get drunk and you do, they'll know. As a parent, you have more influence than you might think.

"Your child is likely to come to you first for information and advice about alcohol, and you can help shape their attitudes and behaviour towards alcohol by reinforcing responsible drinking."

This conversation could include your own alcohol consumption when you were younger: "It's far better to confess, for example, that: 'Yes, I drank at your age – and I wish I hadn't. If I knew then what I know now, I wouldn't have.' And if their questions get uncomfortable, say so."

Children aren't stupid. If you claim you never drink or never get drunk and you do, they'll know."

4. Compare your alcohol consumption to theirs.

Being honest about how much you drink is one thing, but giving children the assumption that they are able to drink the same could cause problems.

Dr Jarvis said: "Young people are going through huge changes in their teens, and in many respects they feel they're grown up.

"You need them to understand that saying they can't drink in the same way you can is nothing to do with you treating them as if they're not mature."

If necessary, it may be helpful for you to explain the physical reasons that children and young people should avoid alcohol at their age.

"Alcohol can harm young people while they are still developing which is why the UK chief medical officers say an alcohol-free childhood is the best option," Dr Jarvis explained.

"Young people's brains are still developing and they may be more vulnerable to long-term effects on memory function, learning ability and educational achievement than adults."

5. Be careful of the language you use.

It's important parents act as good role models regarding alcohol, and this includes how they speak about it in front of their kids.

"What many parents may not realise is that children understand a great deal about the amount they drink and telling stories that glamorise alcohol, can easily undermine other good examples," a spokesperson from Adfam said.

"Parents may be under the impression that stories of their own drunkenness or hangovers may put their children off drinking by highlighting problems, but these stories may have the opposite effect, encouraging and legitimising the idea of excessive drinking."

6. Set rules and boundaries.

"It's also important to set rules surrounding children's drinking," added Dr Jarvis.

"Young people like to push boundaries and test rules. That's part of being a teenager. But the fact is that they feel safer if there are guidelines.

"Have clear rules and have sanctions for breaking them."

A spokesperson from Adfam agreed, adding: "Parents should have open discussions around why these rules are in place.

"Parents who combine warm, two-way conversations and consistent, clear, enforced rules and high supervision, seem best placed to develop secure emotional bonds with their children in a way which could be protective against problematic alcohol use."

13 November 2017

⇨ The above information is reprinted with kind permission from *Huffington Post*. Please visit www. huffingtonpost.co.uk for further information.

Children affected by a parent's drinking

When drinking is a problem

Anthony is already in bed when he hears the front door slam. He covers his head with his pillow to drown out the predictable sounds of his parents arguing. Anthony is all too aware that his father has been drinking again and his mother is angry.

Many teens like Anthony live with a parent who is dependent on alcohol, a person who is physically and emotionally addicted to alcohol. Alcohol dependency has been around for centuries, yet no one has discovered an easy way to prevent or stop it. It continues to cause anguish not only for the person who drinks, but for everyone who is involved with that person.

Why does my parent drink?

Alcohol dependency can be called a disease because the person is at a dis-ease with themselves. Like any disease, it needs to be treated. Without professional help, an alcohol-dependent person will probably continue to drink and may even become worse over time.

Just like any other disease, alcohol dependency is no one's fault. No one sets out to develop this dependency. Some people who live with alcohol-dependent people blame themselves for their loved one's drinking. But the truth is, because of this disease, alcohol-dependent persons would drink anyway. If your parent drinks, it won't change anything if you do better in school, help more around the house, or do any of the other things you may believe your parent wants you to do.

Other people may tell themselves that their parents drink because of some other problem, such as having a rough time at work or being out of work altogether. Parents may be having marital problems, financial problems or someone may be sick. But even if an alcohol-dependent parent has other problems, nothing you can do will make things better. The person with the drinking problem has to take charge of it. No one else can help the dependent person get well.

Why won't my parent stop drinking?

Denial can play a big role in an alcohol-dependent person's life. A person in denial is one who refuses to believe the truth about a situation. A problem drinker may blame another person for the drinking because it is easier than taking responsibility for it. Some alcohol-dependent parents make their kids feel bad by saying things like, "You're driving me crazy!" or "I can't take this anymore."

An alcohol-dependent parent may become enraged at the slightest suggestion that drinking is a problem. Those who acknowledge their drinking may show their denial by saying, "I can stop anytime I want to," "Everyone drinks to unwind sometimes," or "My drinking is not a problem."

Why do I feel so bad?

If you're like most teens, your life is probably filled with emotional ups and downs, regardless of what's happening at home. Add a parent with a drinking problem to this tumultuous time and a person's bound to feel overwhelmed. Teens with alcohol-dependent parents might feel anger, sadness, embarrassment, loneliness, helplessness and a lack of self-esteem.

These emotions can be triggered by the added burdens of living with an alcohol-dependent parent. For example, many alcohol dependents behave unpredictably, and kids who grow up around them may spend a lot of energy trying to feel out a parent's mood or guess what he or she wants. One day you might walk on eggshells to avoid an outburst because the dishes aren't done or the lawn isn't mowed; the next day, you may find yourself comforting a parent who promises that things will be better.

There may be problems paying the bills, having your Mam or Dad show up for important events, and you may even have to take care of younger siblings, too. The pressure to manage these situations in addition to your own life – and maybe take care of younger siblings, too – can leave you exhausted and drained.

Although alcohol dependency causes similar patterns of damage to many families, each situation is unique. Some parents with alcohol problems might abuse their children emotionally or physically. Others neglect their kids by not providing sufficient care and guidance. Parents with alcohol problems may also use other drugs. Your family may have money troubles.

Although each family is different, teens with alcohol-dependent parents almost always report feeling alone, unloved, depressed, or burdened by the secret life they lead at home. Because it's not possible to control the behaviour of an alcohol dependent person, what can a person do to feel better?

What can I do?

Teenage children of alcohol dependents are at a higher risk of becoming alcoholics themselves. Acknowledging the problem and reaching out for support can help ensure that your future does not repeat your parent's past.

Acknowledge the problem

A parent who is a problem drinker is never your fault. Many kids of alcohol-dependent parents try to hide the problem or find themselves telling lies to cover up for a parent's drinking. Admitting that your parent has a problem – even if he or she won't – is the first step in taking control.

Being aware of how your parent's drinking affects you can help put things in perspective. For example, some teens who live with alcohol-dependent adults become afraid to speak out or show any normal anger or emotion because they worry it may trigger a parent's drinking binge.

Clearly, hiding your feelings can create its own set of problems. Acknowledging feelings of anger or resentment – even if it's just to yourself or a close friend – can help protect against this. Recognising the emotions that go with the problem also can help you from burying your feelings and pretending that everything's OK.

Likewise, realising that you are not the cause of a parent's drinking problem can help you feel better about yourself.

Find support

It's good to share your feelings with a friend, but it's equally important to talk to an adult you trust. A school counsellor, favourite teacher or coach may be able to help. Some teens turn to their school, a sympathetic uncle or aunt.

Because alcohol dependency is such a widespread problem, several organisations offer confidential support groups and meetings for people living with alcohol dependents. Al-Anon, an organization designed to help the families and friends of alcoholics, has a group called Alateen that is specifically geared to young people living with adults who have drinking problems. Alateen is not only for children of alcohol-dependent parents, it can also help teens whose parents may already be in recovery. The group Alcoholics Anonymous (AA) also offers a variety of programmes and resources for people living with alcohol dependents.

You're not betraying your parent by seeking help. Keeping 'the secret' is part of the disease of alcohol dependency – and it allows the problems to get worse. As with any disease, it's still possible to love a parent with alcohol dependency while recognising the problems that he or she has. And it's not disloyal to seek help in dealing with the problems your parent's drinking create for you. In fact, taking care of yourself is what your Dad or Mum would want you to do if he or she could think about it clearly!

Find a safe environment

If you find yourself avoiding your house as much as possible, or if you're thinking about running away, consider whether you feel in danger at home. If you feel that the situation at home is becoming dangerous, you can call Childline and never hesitate to dial 999 if you think you or another family member is in immediate danger.

Because alcohol dependency is a disease and not a behaviour, chances are that you won't be able to change your parent's actions. But you can show your love and support – and, above all, take care of yourself.

⇨ The above information is reprinted with kind permission from Drugs and Alcohol Information and Support. Please visit www.drugs.ie for further information

© 2018 Drugs and Alcohol Information and Support

Who will binge drink at age 16? European teen study pinpoints predictors

An international collaboration of scientists leading the world's largest longitudinal adolescent brain imaging study to date has learned that it is possible to predict teenage binge drinking.

The research, published in *Nature*, found that aspects of life experience, personality and brain structure are strong determinants of future alcohol misuse. New simplified versions of the tests are being developed so that children who are at risk of alcohol misuse can be identified and given help.

The data used in the study was collected from the European IMAGEN cohort, led by King's College London, which aims to learn more about biological and environmental factors that might have an influence on the mental health of teenagers.

The study is the first comprehensive analysis of potential influences involved in teenage binge drinking. The researchers used a model which incorporated factors known or believed to be relevant for the development of adolescent substance abuse. These include personality, history/life events, brain physiology and structure, cognitive ability, genetics and demographics – in total 40 different variables were investigated.

Surprisingly, when developing their model to predict teenage binge-drinking, the scientists found that even one to two instances of alcohol consumption by age 14 was sufficient to predict if the teenagers would binge drink at age 16. Previous research has suggested that the odds of adult alcohol dependence can be reduced by 10% for each year that alcohol consumption is delayed in adolescence.

Professor Gunter Schumann, co-author of the paper from the MRC Social, Genetic and Developmental Psychiatry (SGDP) Centre at the Institute of Psychiatry, King's College London, and Coordinator of the IMAGEN project, says: "We aimed to develop a 'gold

standard' model for predicting teenage behaviour which can be used as a benchmark for the development of simpler, widely applicable prediction models. This work will inform the development of specific early interventions in carriers of the risk profile to reduce the incidence of adolescent substance abuse. We now propose to extend analysis of the IMAGEN data in order to investigate the development of substance use patterns in the context of moderating environmental factors, such as exposure to nicotine or drugs as well as psychosocial stress."

IMAGEN recruited over 2,000 teenagers from England, Ireland, France and Germany at age 14 years. Follow-up work at age 16, funded by the Medical Research Council (MRC), has shown that it is possible to predict future alcohol misuse two years later, and the scientists wish to continue this work by re-assessing the participants at a later age. The factors assessed in this study will also be applied to predict other types of risk-taking behaviours, such as drug-taking and smoking.

Early onset of teenage binge drinking and progression to alcohol misuse has previously been shown to be genetically influenced and has been consistently shown to be associated with later risk for substance use disorders. However, it is important to understand whether environmental factors can modify the risk imposed by our genes. In this study, negative life experiences were shown to be an important influence on binge drinking behaviour at the age of 14.

Dr Robert Whelan, lead author of the study, formerly from the University of Vermont and currently at University College Dublin, says: "Our goal was

to better understand the relative roles of brain structure and function, personality, environmental influences and genetics in the development of adolescent abuse of alcohol. This multidimensional risk profile of genes, brain function and environmental influences can help in the prediction of binge drinking at age 16 years."

Hugh Perry, chair of the MRC Neurosciences and Mental Health Board, says: "Addiction and substance misuse is a major medical, social and economic problem for the UK. The UK Government spends more than £15 billion annually in meeting the cost of drug-related social and economic harm. The MRC is supporting research that aims to identify the medical harms caused by alcohol consumption and linking these to the various drinking behaviours prevalent in the UK. We believe that establishing such links will lead to breakthroughs in this field and provide compelling evidence to inform public health policy and lay the groundwork for the design of interventions."

3 July 2014

⇨ The above information is reprinted with kind permission from the Institute of Psychiatry, Psychology & Neuroscience (IoPPN), King's College, London. Please visit www. kcl.ac.uk for further information.

Parents of 'socially advantaged children' most likely to allow kids to drink by age 14

"They appear to view alcohol use as less risky."

By Amy Packham

By the age of 14, almost half of kids have tried more than a few sips of alcohol and 'socially-advantaged' children were the group most likely to drink.

An analysis of 10,000 children born in the UK, which is part of the Millennium Cohort Study, found that one in six parents allow their kids to drink booze by the time they turn 14.

The researchers, from the Centre for Longitudinal Studies at the UCL Institute of Education and Pennsylvania State University, found that "well-educated parents of white children" were most likely to allow their children to drink at 14.

Some parents may believe that exposing their kids to alcohol will help them develop a healthy attitude to drinking, but the study's lead author, Jennifer Maggs, advised against this.

"Parents of socially-advantaged children may believe that allowing children to drink will teach them responsible use or may in fact inoculate them against dangerous drinking... they appear to view alcohol use as less risky," said Maggs.

"However, there is little research to support these ideas."

For the study, researchers also looked at parents' drinking habits and attitudes to drinking.

The results showed that mums and dads who were "light or moderate" drinkers were just as likely to let their children drink as those who drank heavily.

Parents who did not drink alcohol were less likely to allow their kids to drink.

According to the Institute of Alcohol Studies, (IAS) 2017 report, English Chief Medical Officers recommend a child does not drink before the age 15, and between the ages of 15 and 18 drinking should be supervised by an adult.

Katherine Brown, chief executive of the Institute of Alcohol Studies, told *The Guardian*: "It is worrying to see that this advice may not be getting across to parents, who are trying to do their best to teach their children about alcohol. We need to see better guidance offered to parents via social marketing campaigns and advice from doctors and schools."

Dr Sarah Jarvis, medical advisor to Drinkaware, previously told HuffPost

UK many parents may not be aware how alcohol can affect young people.

"Alcohol can harm young people while they are still developing which is why the UK chief medical officers say an alcohol-free childhood as the best option," Dr Jarvis explained.

"Young people's brains are still developing and they may be more vulnerable to long-term effects on memory function, learning ability and educational achievement than adults."

The latest figures from NHS show 44% of 11- to 15-year-olds have tried alcohol. The NHS states that children and young people are advised not to drink alcohol before the age of 18.

"Alcohol use during the teenage years is related to a wide range of health and social problems," they stated. "However, if children do drink alcohol underage, it shouldn't be until they are at least 15."

The law states it is illegal to give children alcohol if they are under five, therefore it is not illegal for a child aged five to 16 to drink alcohol at home or on other private premises.

15 December 2017

⇨ The above information is reprinted with kind permission from *Huffington Post*. Please visit www. huffingtonpost.co.uk for further information.

© 2018 AOL (UK) Limited

On children of alcoholics: a manifesto for change

An extract from an article by the All-Party Parliamentary Group on children of alcoholics.

One in five children in the UK lives with a parent who drinks too much – that's over 2.5 million children. They are Britain's innocent victims of drink. Hard-drinking parents hurt their children for life. Compared to other children, children of alcoholics are:

⇨ twice as likely to experience difficulties at school;

⇨ three times more likely to consider suicide; and

⇨ five times more likely to develop eating disorders.

Worst of all, children of alcoholics are also four times more likely to become alcoholics themselves – there is a cycle of alcoholism cascading down the generations. We have to break the cycle of this terrible disease – and that starts by breaking the silence around Britain's biggest secret scandal. The APPG's research confirms a shocking picture of support for children of alcoholics:

1. None of the 138 respondent Local Authorities have a specific strategy for support for children of alcoholics.

2. Almost no Local Authority is increasing its drug and substance abuse treatment budgets, despite the increases in alcohol-related hospital admissions. Of the 49 Local Authorities providing data on future treatment budgets, 70% (34 Local Authorities) are experiencing rising alcohol-related hospital admissions.

⇨ Yet only 9% of these Local Authorities are increasing treatment budgets (three Local Authorities in total).

⇨ Over a third are actually cutting treatment budgets (12 Local Authorities).

3. The number of people accessing alcohol treatment varies widely, from 0.4% of a Local Authority's estimated number of hazardous drinkers to 11%.

4. There is huge variation in average drug and substance abuse treatment budgets for hazardous drinkers – from £6.61 a head on the Isle of Wight to £419.04 in Sefton.

5. There is very little uniformity in the data provided by different authorities. Although a number of national measurement systems for alcohol misuse are available, these are not used by all Local Authorities.

The hidden stigma attached to children of alcoholics typically means that they suffer in silence. This needs to change. Over the last year, the All-Party Parliamentary Group on Children of Alcoholics has brought together policy makers, experts from charities, interest groups and medicine and – most importantly – children of alcoholics themselves, to spell out what government can start doing and do better.

This manifesto sets out the ten key points the Government needs to address if children of alcoholics are to be properly supported and the shocking rise in problem drinking halted.

The Government needs to

1. Take responsibility for children of alcoholics

2. Create a national strategy for children of alcoholics

3. Properly fund local support for children of alcoholics

4. Increase availability of support for families battling alcohol problems

5. Boost education and awareness for children

6. Boost education and training for professionals with a responsibility for children

7. Develop a plan to change public attitudes

8. Revise the national strategy to tackle alcohol harm to focus on price and availability

9. Curtail the promotion of alcohol – especially to children

10. Take responsibility for reducing rates of alcohol harm

Government needs to take responsibility for children of alcoholics

The Government should take responsibility for supporting children of alcoholics. They are Britain's innocent victims of drink.

Children of alcoholics are currently a forgotten part of the Government's stance on alcohol.

This means children of alcoholics should be properly recognised within existing alcohol policy and mental health services.

The Government needs to pay more attention to the issue of children of alcoholics but by providing support facilities for those children alongside treatment facilities for existing alcoholics to reduce the prevalence of alcoholism and children of alcoholics in this generation and instances of parents drinking too heavily in future generations.

Create a national strategy for children of alcoholics

The Government needs to introduce a national strategy for children of alcoholics.

The challenge is that children of alcoholics sit on the fault-line of three different systems and are falling through the gaps between:

⇨ the adult social care system, which might target help at adults but only really those adults with acute needs;

⇨ the children's social care system, which can provide support but, again, only really if the need is acute; and

⇨ the public health system, which is yet to properly develop effective ways of reaching out and helping children of alcoholics specifically.

This strategy will only work if it includes:

1. Appointing a Minister to take charge of policy for children of alcoholics and coordinating policy across government; this brief could be included in that of the Minister of State for Vulnerable Children and Families, for example.

2. Creating a new obligation on Local Authorities and the NHS in every part of the country to:

⇨ tell us the scale of the challenge in their area, including an estimate of the number of children of alcoholics in their area, using standardised definitions and methods of collecting information;

⇨ report every year on what they are doing to support children of alcoholics; and report every year on what they are spending.

In addition, the strategy needs to provide:

⇨ properly funded and resourced support facilities for children of alcoholics;

⇨ a plan to deliver better awareness among children of alcoholics, including education about the effects of alcohol and what to do if they have heavy-drinking parents; and

⇨ proper investment in treatment facilities for parents who drink too much.

Crucially, there needs to be a designated person within government, at the central and local level, with responsibility for this strategy and for children of alcoholics as a whole.

⇨ The above extract is reprinted with kind permission from appg on children of alcholics. Please visit www.nacoa.de for further information.

© 2018 The All-Party Parliamentary Group

Generation Y have turned their Backs on alcohol

Exclusive Demos polling for *Character and Moderation: Alcohol* ***paper finds health and money-conscious Millennials are shunning alcohol at unprecedented levels.***

19 per cent of 16–24s don't drink, and 66 per cent don't feel alcohol is important to their social lives.

41 per cent of Gen-Y drinkers think alcohol is less important to their own social life than to their parents'.

Young people cite growing awareness of health, not being able to afford alcohol, and alcohol being harder to get hold of as key drivers behind their reduced drinking habits.

The findings suggest a seismic cultural shift in youth drinking habits, and validates official statistics showing young people are drinking less than their counterparts were ten years ago.

Exclusive new research from Demos think tank has found that the vast majority of 16–24s either don't drink (19 per cent) or feel alcohol isn't important to their social lives (66 per cent).

The polling of 16–24-year-olds confirms ONS statistics, which have shown a marked decline in self-reporting of youth drinking habits over the past decade.

It also reveals that some young people think that alcohol is more important to their parents' lives than to their own (30 per cent).

While it has been speculated that an increase in migrant populations from non-drinking cultures could stand behind the falling alcohol rates, Demos' own analysis shows this would only account for up to 31 per cent of the rise in the number of teetotallers.

The polling rather suggests that the declining consumption levels represent both a substantial cultural shift amongst young people, and positive progress towards a policy ambition of successive governments, with 66 per cent citing increased awareness in the health consequences of excessive drinking as contributing to the fall.

Other popular factors included young people being less able to afford alcohol compared to ten years ago (55 per cent) and many believing that alcohol is now more difficult to obtain under-age compared to ten years ago (47 per cent). Over 40 per cent (42 per cent) also cited the time young people spend on social media and the Internet as having contributed to the decline in alcohol consumption.

Against the backdrop of positive change in young people's drinking habits overall, however, *Character and Moderation: Alcohol* highlights how serious problems with youth drinking remain, particularly in certain areas of the country, and that young people with a history of alcohol abuse in their family continue to remain particularly vulnerable to developing unhealthy relationships with alcohol.

To build on the recent progress and target those who remain most at risk, the paper recommends both local and national governments, public health organisations and the alcohol industry take decisive action, through the following recommendations.

Central Government should:

1. Provide a comprehensive early intervention strategy as part of its strategy to tackle alcohol misuse.

2. Continue to target resources at the home environment and support for parents, particularly those in vulnerable situations, through increased investment in Family Nurse Partnerships.

3. Link the size of public health budgets that local authorities receive to alcohol harm profiles.

4. Encourage improved joined-up working between government departments with current responsibility for alcohol (Home Office, Department for Health, Public Health England), the Department for Education and the Cabinet Office.

The Department for Education should:

5. Ensure that teacher training colleges are teaching best practice pedagogical approaches to ensure that teachers adopt teaching strategies that evidence shows are more likely to build character in their pupils.

6. Embed Personal, Social and Health Education (PSHE) within the national curriculum, and incentivise schools to adopt a 'whole school' approach to character development.

Public Health England should

7. Work with local authorities and the Department for Education to ensure that 'life skills' programmes in schools are considered an important component of public health strategies at a local level.

8. Invest in further research to understand what is causing the sustained decline in youth drinking.

Local Governments should

9. Help to strengthen local alcohol partnerships to curb underage drinking – working with schools and public health workers – and continue to promote diversionary activities and innovations such as non-drinking pubs for young people.

The Alcohol Industry should

10. Look at ways to engage positively with national campaigns aimed at building character skills and healthy lifestyle choices amongst young people.

Commenting on the findings, the paper's author and Head of Citizenship at Demos, Jonathan Birdwell, said:

"These findings strengthen our understanding of a phenomenon that has taken many of us by surprise. They reveal the potential of public policy to both encourage and complement cultural changes, to make a real difference in challenging harmful behaviour. But while the trends are pointing in a positive direction, we cannot ignore the fact that there is still a relatively significant minority of young people indulging in hazardous binge drinking – which is damaging to their health, their career prospects and to society as a whole. It is important for us now to build on these insights and determine the best means of directing limited public funds to tackle this pernicious issue at the root cause."

15 July 2015

⇨ The above extract is reprinted with kind permission from Demos. Please visit www.demos.co.uk for further information.

Energy drinks and alcohol, a risky mix… psychologically

An article from The Conversation.

THE CONVERSATION

Pierre Chandon, Professeur de Marketing et Directeur du Centre Multidisciplaire des Sciences Comportementales, INSEAD; Anna Krishna, Dwight F Benton Professor of Marketing, University of Michigan; and Yaan Cornil, Assistant Professor, Marketing and Behavioural Science Division, Saunder School of Business, University of British Columbia

People who add energy drinks to alcohol have a higher risk of injury from car accidents and fights, compared to those who drink alcohol straight. This is the conclusion of a meta-analysis of 13 studies published in March in the *Journal of Studies on Alcohol and Drugs*.

Energy drinks, such as Red Bull, Monster, or Rockstar, contain ingredients that are considered stimulants, such as caffeine or guarana. The reality of their effects has however been a matter of dispute for several years already.

An experiment conducted by our team from INSEAD Business School, the University of British Columbia, and the University of Michigan, shed new lights on the effects of mixing alcohol and energy drinks. In an article forthcoming in the *Journal of Consumer Psychology*, already available online, we show that the associations that people have with the popular vodka and Red Bull cocktail can increase perceived intoxication and lead to risky behaviours.

However, these effects are not driven by the ingredients contained in energy drinks. They are linked to the beliefs that people have that energy drinks boost the intoxicating effects of alcohol. It is a psychological effect, not a physiological one.

73 per cent of US college students mix alcohol and energy drinks

Cocktails mixing alcohol and energy drinks are popular in many countries. A 2011 study in a French university found that they were consumed by 54 per cent of French students (64 per cent for males and 46 per cent for females). This proportion reached 73 per cent among American students and 85 per cent in Italy, according to a study published in the *British Medical Journal*.

Compared to people who drink alcohol straight, those who mix it with energy drinks have double the risk of experiencing or committing sexual assault, or of being involved in a drunk driving accident, according to a *JAMA* article.

Some researchers have hypothesised the existence of a causal relation between the ingredients present in energy drinks and risky behaviours.

They argue that energy drinks, because of the caffeine that they contain, mask the intoxication effects of alcohol, fooling people who are drunk into believing that they are not.

The masking hypothesis, however, has been refuted in a recent meta-analysis of nine studies. This study showed that that the amount of caffeine typically found in energy drinks is too low to change perceived intoxication.

The role of beliefs

All these studies have in common that they were 'blind,' meaning that the participants did not know whether they were consuming alcohol mixed with an energy drink or alcohol alone. Because of that, these studies missed an important part of the story: the psychological impact that energy drinks can have because of people's beliefs.

For our experiment, we recruited 154 young heterosexual Parisian men of comparable body mass who were social drinkers but had no risk of alcohol dependence. Under the pretence of studying bar behaviours, we invited them to the INSEAD Sorbonne University Behavioural Lab, in Paris. We asked them to drink a cocktail containing six centilitres (two ounces) of 40 per cent Smirnoff vodka (a common amount in one drink), eight centilitres (2.7 ounces) of Red Bull Silver Edition energy drink, and 16 centilitres (5.4 ounces) of Caraïbos Nectar Planteur (a blend of fruit juices).

We randomly assigned the participants to one of three groups, where the only difference was the name the drink was called by – 'vodka Red Bull cocktail' which emphasised both the energy drink and the alcohol, 'vodka cocktail' which emphasised only the alcohol, and 'exotic fruits cocktail' which emphasised neither alcohol nor the energy drink.

Measuring sexual confidence and risky behaviour

After waiting for 30 minutes for the cocktail to have an effect, we showed the (male, heterosexual) participants, photos of 15 young women, one by one. After looking at each photo, the participants reported their intention to approach and 'chat up' the woman represented in each photo, and their prediction of whether the woman would share her phone number. Measures such as these were used to create sexual self-confidence scores.

To measure general risk-taking, participants could earn money by

blowing up a virtual balloon. Each pump inflated the balloon and added money to a counter. Participants could cash-out before the balloon exploded (which happened randomly) or keep pumping at the risk that it would explode, resulting in the loss of the money accumulated on the trial.

Finally, we also asked the participants how long they would wait (in minutes) to sober up before driving, and how drunk they felt.

Same cocktail, different sensations of drunkenness...

Given that the young males in all three groups had the same drink, there was no difference in actual intoxication levels across the three drink-name groups. However, people in these three groups felt drunk to different extents – those in the 'vodka Red Bull' label group felt 51 per cent more drunk than those in the other two groups. This effect was even stronger for participants who most strongly believe that energy drinks increase alcohol intoxication. This is characteristic of a placebo effect, like the those found among patients taking an innocuous drug. So, the more people believe that energy drinks boost the effects of alcohol, the more the 'vodka Red Bull' label increased perceived intoxication.

Next, we found that the young men in the 'vodka Red Bull' label group took more risks in the balloon game and

were more sexually self-confident. These placebo effects were also reinforced by how much people believed that drunkenness increases impulsiveness and removes sexual inhibitions. The silver lining in the study was that that the 'vodka Red Bull' label also made participants intend to wait longer before driving.

Our results suggest a causal relationship between mixing energy drinks and alcohol and two risky behaviours, seduction behaviours and gambling. They confirm the association found in the meta-analysis mentioned earlier but contribute by providing evidence for one mechanism for the effect, albeit a psychological, not a physiological, one.

...and varying caution before driving

One of the unexpected results of our study is that emphasising the presence of an energy drink in an alcohol cocktail makes people more careful before considering driving. This may seem to be contradicted by the association between car crashes and the consumption of alcohol mixed with energy drinks. It suggests that this association may be explained, not by a cause-and-effect relation, but by the possible fondness of those who drive recklessly for mixing alcohol and energy drinks. It is a hypothesis that merits further research.

The key issue is the promotion of disinhibiting and risky behaviours by energy-drink brands through sponsorship and advertising. We now know that exposing consumers to these messages turns an innocuous ingredient into an active placebo. Even if these effects are only fantasised, their consequences are nonetheless real.

11 May 2017

⇨ The above information is reprinted with kind permission from *The Conversation*. Please visit www.theconversation.com for further information.

Alcohol

Getting the facts

Just about everyone knows that the legal drinking age throughout the United States is 21. But according to the National Center on Addiction and Substance Abuse, almost 80% of high school students have tried alcohol.

Deciding whether to drink is a personal decision that we each eventually have to make. This article provides some information on alcohol, including how it affects your body, so you can make an educated choice.

What is alcohol?

Alcohol is created when grains, fruits or vegetables are fermented. Fermentation is a process that uses yeast or bacteria to change the sugars in the food into alcohol. Fermentation is used to produce many necessary items – everything from cheese to medications. Alcohol has different forms and can be used as a cleaner, an antiseptic, or a sedative.

So if alcohol is a natural product, why do teens need to be concerned about drinking it? When people drink alcohol, it's absorbed into their bloodstream. From there, it affects the central nervous system (the brain and spinal cord), which controls virtually all body functions. Because experts now know that the human brain is still developing during our teens, scientists are researching the effects drinking alcohol can have on the teen brain.

How does it affect the body?

Alcohol is a depressant, which means it slows the function of the central nervous system. Alcohol actually blocks some of the messages trying to get to the brain. This alters a person's perceptions, emotions, movement, vision and hearing.

In very small amounts, alcohol can help a person feel more relaxed or less anxious. More alcohol causes greater changes in the brain, resulting in intoxication. People who have overused alcohol may stagger, lose their coordination and slur their speech. They will probably be confused and disoriented. Depending on the person, intoxication can make someone very friendly and talkative or very aggressive and angry. Reaction times are slowed dramatically – which is why people are told not to drink and drive. People who are intoxicated may think they're moving properly when they're not. They may act totally out of character.

When large amounts of alcohol are consumed in a short period of time, alcohol poisoning can result. Alcohol poisoning is exactly what it sounds like – the body has become poisoned by large amounts of alcohol. Violent vomiting is usually the first symptom of alcohol poisoning. Extreme sleepiness, unconsciousness, difficulty breathing, dangerously low blood sugar, seizures, and even death may result.

Why do teens drink?

Experimentation with alcohol during the teen years is common. Some reasons that teens use alcohol and other drugs are:

⇨ curiosity

⇨ to feel good, reduce stress and relax

⇨ to fit in

⇨ to feel older.

From a very young age, kids see advertising messages showing beautiful people enjoying life – and alcohol. And because many parents and other adults use alcohol socially – having beer or wine with dinner, for example – alcohol seems harmless to many teens.

Why shouldn't I drink?

Although it's illegal to buy alcohol in the United States until the age of 21, most teens can get access to it. It's therefore up to you to make a decision about drinking. In addition to the possibility of becoming addicted, there are some downsides to drinking:

The punishment is severe. Teens who drink put themselves at risk for obvious problems with the law (it's illegal; you can get arrested). Teens who drink are also more likely to get into fights and commit crimes than those who don't.

People who drink regularly also often have problems with school. Drinking can damage a student's ability to study well and get decent grades, as well as affect sports performance (the coordination thing).

You can look really stupid. The impression is that drinking is cool, but the nervous system changes that come from drinking alcohol can make people do stupid or embarrassing things, like throwing up or peeing on themselves. Drinking also gives people bad breath, and no one enjoys a hangover.

Alcohol puts your health at risk. Teens who drink are more likely to be sexually active and to have unsafe, unprotected sex. Resulting pregnancies and sexually transmitted diseases can change – or even end – lives. The risk of injuring yourself, maybe even fatally, is higher when you're under the influence, too. One half of all drowning deaths among teen guys are related to alcohol use. Use of alcohol greatly increases the chance that a teen will be involved in a car crash, homicide or suicide.

Teen drinkers are more likely to get fat or have health problems, too. One study by the University of Washington found that people who regularly had five or more drinks in a row starting at age 13 were much more likely to be overweight or have high blood pressure by age 24 than their nondrinking peers. People who continue drinking heavily well into adulthood risk damaging their organs, such as the liver, heart, and brain.

How can I avoid drinking?

If all your friends drink and you don't want to, it can be hard to say "no, thanks". No one wants to risk feeling rejected or left out. Different strategies for turning down alcohol work for different people. Some people find it helps to say no without giving an explanation, others think offering their reasons works better ("I'm not into drinking", "I have a game tomorrow", or "my uncle died from drinking", for example).

If saying no to alcohol makes you feel uncomfortable in front of people you know, blame your parents or another adult for your refusal. Saying, "My parents are coming to pick me up soon", "I already got in major trouble for drinking once, I can't do it again", or "my coach would kill me", can make saying no a bit easier for some.

If you're going to a party and you know there will be alcohol, plan your strategy in advance. You and a friend can develop a signal for when it's time to leave, for example. You can also make sure that you have plans to do something besides just hanging out in someone's basement drinking beer all night. Plan a trip to the movies, the mall, a concert or a sports event. You might also organise your friends into a volleyball, bowling, or softball team – any activity that gets you moving.

Girls or guys who have strong self-esteem are less likely to become problem drinkers than people with low self-esteem.

Where can I get help?

If you think you have a drinking problem, get help as soon as possible. The best approach is to talk to an adult you trust. If you can't approach your parents, talk to your doctor, school counsellor, clergy member, aunt or uncle. It can be hard for some people to talk to adults about these issues, but a supportive person in a position to help can refer students to a drug and alcohol counsellor for evaluation and treatment.

In some states, this treatment is completely confidential. After assessing a teen's problem, a counsellor may recommend a brief stay in rehab or outpatient treatment. These treatment centres help a person gradually overcome the physical and psychological dependence on alcohol.

What if I'm concerned about someone else's drinking?

Sometimes people live in homes where a parent or other family member drinks too much. This may make you angry, scared, and depressed. Many people can't control their drinking without help. This doesn't mean that they love or care about you any less. Alcoholism is an illness that needs to be treated just like other illnesses.

People with drinking problems can't stop drinking until they are ready to admit they have a problem and get help. This can leave family members and loved ones feeling helpless. The good news is there are many places to turn for help: a supportive adult, such as your guidance counsellor, or a relative or older sibling will understand what you're going through. Also, professional organisations like Alateen can help.

If you have a friend whose drinking concerns you, make sure he or she stays safe. Don't let your friend drink and drive, for example. If you can, try to keep friends who have been drinking from doing anything dangerous, such as trying to walk home at night alone or starting a fight. And protect yourself, too. Don't get in a car with someone who's been drinking, even if that person is your ride home. Ask a sober adult to drive you instead or call a cab.

Everyone makes decisions about whether to drink and how much – even adults. It's possible to enjoy a party or other event just as much, if not more so, when you don't drink. And with your central nervous system working as it's supposed to, you'll remember more about the great time you had!

September 2016

The above information is reprinted with kind permission from The Nemours Foundation/KidsHealth. Please visit www.kidshealth.org for further information.

Young turn their back on drink as teetotalism flourishes

Total abstinence in fashion among the young – but elderly women are drinking more.

Teetotalism has become a major force in British life for the first time since the industrial heyday of the temperance movement after a dramatic rise in the number young people shunning alcohol.

The number of under-25s opting for total abstinence from drink has leapt by 40 per cent in just eight years as young people overtake the elderly as the most sober generation.

New figures published by the Office for National Statistics show that more than a quarter of young people do not drink alcohol at all and binge drinking is also in decline.

In London, which is both the youngest and most ethnically diverse region of the UK, one in three people is now teetotal.

Once closely associated with nonconformist Christian groups such Methodists and the Salvation Army, the rise in total abstinence has coincided with the growth of the Muslim community in Britain.

But Islam is likely only to account for a small part of the phenomenon amid a wider shift in attitudes among the young.

It follows a series of studies suggesting that the generation which came of age in the era of austerity and university tuition fees also hold more conservative views on drinking, smoking, gambling and sex than their predecessors.

Binge drinking – measured as the number of people who had a heavy drinking session in the week before the annual ONS General Household Survey – fell by almost 17 per cent across the wider population between 2005 and 2013. But among young people it dropped by a third.

Only one in 50 men under 25 now drink on an almost daily basis – compared with one in ten a decade ago.

While in 2005 elderly people were a third more likely than the young to abstain from alcohol, today the gap has disappeared altogether, with 27 per cent of both age groups now teetotal.

Although women are still significantly more likely than men to abstain from alcohol, the gap has narrowed.

Strikingly, the number of men aged between their mid-40s and mid-60s turning their back on drink altogether has jumped by 38 per cent since 2005 – but among women of the same age it fell by seven per cent.

And among retired women the pattern in drinking habits is effectively the reverse of that seen across the rest of the population.

The number of women over 65 abstaining from alcohol fell by 16 per cent in relative terms between 2005 and 2013.

London, had by far the highest prevalence of teetotalism in the study, which covered mainland Britain, with 32 per cent abstaining.

The West Midlands was second with 25 per cent while the North East and South West regions of England had the lowest abstinence levels, at 17 per cent and 15 per cent, respectively.

"Although it is difficult to attribute regional differences to any single factor, London is the most ethnically diverse region of the UK and has a lower than average population age of just 33," the ONS noted.

"Both of these factors may play a part in London having a higher than average number of teetotallers."

It added that changing attitudes among the young could be a mark of the success of campaigns to clamp down on underage drinking when they were children.

"It is known that people who start drinking at a younger age are likely to drink more frequently, and in greater quantities, in adulthood," the ONS remarked.

Professor Sir Ian Gilmore, chairman of the Alcohol Health Alliance UK said: "These results today are an encouraging sign, however there is absolutely no room for complacency.

"Whilst the average level of consumption has fallen, this may be in part due to the change in the ethnic make-up of the country with many people abstaining from drinking altogether.

"Data on alcohol consumption is also unreliable; many people under-report how much they drink and self-reported data on consumption does not correlate with sales data.

"Enough alcohol is sold in the UK for all consumers to be drinking above the low-risk guidelines. Alcohol remains the biggest single cause of death in under-60s in the UK and globally, more alarmingly, people are dying at a far younger age.

"The overall numbers of alcohol-related deaths may be down but the numbers are still far higher than they were 20 years ago. Without effective action from government on pricing, marketing and availability, we are storing up major problems for the future."

13 February 2015

⇨ The above information is reprinted with kind permission from *The Telegraph*. Please visit www. thetelegraph.co.uk for further information.

Alcohol advertising has no place on our kids' screens

THE CONVERSATION

An article from **The Conversation.**

By Sandra Jones, Professor and Director of the Centre for Health and Social Research, Australian Catholic University

Research shows children who are exposed to alcohol advertising are more likely to start drinking earlier and to drink more.

But Australian regulations are inadequate to protect children and adolescents from such advertising. That's the finding of the Australian National Preventive Health Agency (ANPHA) review of alcohol advertising.

Importantly, it makes 30 recommendations to fix the problem.

But 18 months after ANPHA passed the report to government, the Coalition is yet to formally release it, or act on the recommendations. The Foundation for Alcohol Research & Education (FARE) obtained the report under FOI and posted it online today.

Placement of alcohol advertising

Research from the Unites States shows sixth graders' exposure to outdoor ads such as billboards and bus stops predicts their alcohol attitudes and intentions at eighth grade.

In Australia, the Outdoor Media Association requires its members not to advertise alcohol on billboards or fixed signs within 150 metres of a primary or secondary school, except where the school is near a club, pub or bottle shop.

ANPHA describes this as a token gesture. It recommends:

⇨ the distance be increased to 500m

⇨ loopholes, such as the exemption for retail ads, be closed

⇨ a system be established to monitor compliance with the rules.

Free TV Australia allows alcohol ads only during periods of M, MA or AV programming, except during the live broadcast of sporting events on weekends and public holidays.

Research released last month found that, due to the sporting "loophole", children aged under 18 years were exposed to a cumulative total of 51 million alcohol ads in 2012.

ANPHA recommends the code of practice be revised to remove the exclusions for sports broadcasts and school days between 12pm and 3pm.

The report notes that children are increasingly exposed to alcohol advertising on the Internet.

Our research found that Australian alcohol brand websites typically had poor filter systems for preventing access by children. Only half required the user to provide a date of birth. None had any means of preventing users from trying again.

ANPHA recommends that:

⇨ alcohol companies use adequate age checks on their own websites

⇨ social media providers (such as YouTube and Facebook) implement age-gating tools to reduce access by children.

The report notes there are currently no restrictions on sponsorship of sport by alcohol companies in Australia. Children are therefore increasingly exposed to alcohol brands in sporting environments.

Researchers in Western Australia found that children as young as ten can recall which sponsors are associated with sporting teams. And our research found that children associate these products with sport and with positive personal characteristics and outcomes.

The report also recommends restrictions on alcohol-branded merchandise and on alcohol advertising on subscription television and in cinemas between 5am and 8.30pm.

Content of alcohol advertising

The content of alcohol advertising is covered by the Alcohol Beverages Advertising Code (ABAC) Scheme. The report notes that ministerial reviews of the scheme in 2003 and 2009 found that:

⇨ the system of self-regulation of alcohol advertising exhibited serious deficiencies and required much improvement.

Problems with the system include the narrow interpretation of the code in the adjudication of complaints, the lack of clear provisions regarding advertising that appeals to children, the lack of monitoring of compliance with the code, and the absence of penalties for breaching the code.

The report notes that there have been several improvements following the reviews, but makes a total of 15 recommendations in relation to the scheme, including improving the review processes.

There is substantial research evidence to show the code is ineffective. We, like many others, reported on some of the problems with the system before the 2003 review of the ABAC Scheme; and after the revisions made in response to the review.

Improving transparency

So, why is the Government sitting on the report?

There may be a number of legitimate reasons behind the Government's failure to release the report. Perhaps they forgot. Perhaps they were too busy.

However, the sceptic in me wonders whether part of the reason relates to recommendations 11a and 11b:

⇨ 11a recommends monitoring and reporting on children's and adolescents' exposure to alcohol advertising and the effectiveness

(or otherwise) of current measures to reduce this exposure.

⇨ 11b recommends the Government review progress against the recommendations in 2015. If the recommendations regarding the ABAC Scheme have not progressed and/or little headway has been made on removing the live sports broadcast exemption, "then governments should proceed to legislate to control alcohol advertising and marketing".

Perhaps the delay will give the industry an opportunity to address some of the concerns raised in the report.

However, proposed revisions to the Free TV Code would increase children's exposure by allowing alcohol advertising from the earlier time of 7.30pm.

Based on past evidence, it's not surprising that our government is

reluctant to step in and regulate alcohol advertising.

It's not surprising that our government is reluctant to release a report that recommends greater controls on alcohol advertising.

And it's not surprising that our government continues to allow self-regulatory and quasi-regulatory codes that serve to protect the alcohol industry and fail to protect our children and adolescents.

23 October 2013

⇨ The above information is reprinted with kind permission from *The Conversation*. Please visit www.theconversation.com for further information.

© 2010-2018 The Conversation Trust (UK)

Foul play? New report highlights how alcohol industry bent the rules on advertising during UEFA Euro 2016

A new report highlights how alcohol producers worked to circumvent legislation designed to protect children during the UEFA Euro 2016 football tournament.

Researchers at the Institute for Social Marketing, University of Stirling, found over 100 alcohol marketing references per televised match programme in three countries – France, the UK and Ireland.

Foul play

Most marketing appeared in highly visible places, such as pitch-side advertising during the matches. This was the case, despite the fact that the tournament was held in France, where alcohol TV advertising and sports sponsorship is banned under the 'Loi Évin'.

The report, *Foul Play? Alcohol marketing during UEFA Euro 2016*, will be launched at the European Healthy Stadia conference at Emirates Stadium on Thursday 27 April.

An analysis of broadcast footage found that alcohol marketing appeared, on average, once every other minute. The majority took the form of 'alibi' marketing, whereby indirect brand references are used to promote a product, rather than a conventional logo or brand name. Carlsberg was the most featured brand, accounting for almost all references in each of the three countries, using their slogan 'Probably the best in the world' while avoiding the mentioning the product name. 'Alibi' marketing was a common practice of tobacco companies in sporting events when advertising restrictions were introduced.

Protecting children

Dr Richard Purves, Principal Investigator, Institute for Social Marketing, University of Stirling said:

"Beamed to audiences across the world, major sporting events such as the UEFA EURO tournament, present a prime opportunity for alcohol companies to market directly to a global audience. In order to continue to protect children and young people from exposure to alcohol marketing, laws such as those in France need to be upheld and respected by all parties involved and not seen as something to be negotiated."

Katherine Brown, Director of the Institute of Alcohol Studies said:

"There is strong evidence that exposure to alcohol marketing encourages children to drink earlier and in greater quantities. The findings of this report show that alcohol companies are following in the footsteps of their tobacco colleagues by bending the rules on marketing restrictions putting children's health at risk."

Eric Carlin, Director of Scottish Health Action on Alcohol Problems (SHAAP), said:

"Sport should be an alcohol-free space. The presence of alcohol marketing during UEFA EURO 2016 highlights that organisers of sporting events need to hold out against tactics of big alcohol companies to flout legal regulations designed to protect children."

27 April 2017

⇨ The above information is reprinted with kind permission from The University of Stirling. Please visit www.stir.ac.uk for further information.

Health leaders call for ban on alcohol firms sponsoring sports clubs and events

Health campaigners fear it could fuel underage drinking.

By Heather Saul

Health leaders are calling for alcohol sponsorship in sport to be banned over fears it could fuel underage drinking.

Medical leaders, public health campaigners and health charities signed a letter to *The Guardian* calling on for an end to the "outrageous" practice of alcohol sponsorship in sport.

The signatories urged the Government to "listen to the people rather than to big business".

"Self-regulation of alcohol advertising isn't working when it allows drink brands to dominate sporting events that attract children as well as adults, creating automatic associations between alcohol brands and sport that are cumulative, unconscious and built up over years," the letter stated.

This year's World Cup featured one example of alcohol advertising for each minute of football, the letter's signatories said.

They said people would consider it "outrageous" if tobacco firms became brand ambassadors for big football teams, and questioned the acceptability of drinks advertising in comparison.

The call for the ban comes as thousands across the UK prepare to watch sporting events on Boxing Day, where alcoholic drinks are often promoted.

Its signatories claimed alcohol sponsorship of sport has become "as

commonplace as advertising for cereal or soap powder".

The letter adds: "Let's take action to protect our children by ensuring that the sports we watch promote healthy lifestyles and inspire participation, not a drinking culture. Let's make alcohol sports sponsorship a thing of the past."

A spokeswoman for the Portman group, which represents alcohol producers, told *The Guardian*: "Calling for a ban does not reflect the reality of what is happening in the UK, where official government statistics show that rates of binge drinking among 16- to 24-year-olds are in significant decline and the number of children even trying alcohol is at a record low."

She added that alcohol sponsorship contributed significantly to the country's economy.

26 December 2014

⇨ The above information is reprinted with kind permission from *The Independent*. Please visit www.independent.co.uk for further information.

More Evidence That Alcohol Ads Are Linked To Teen Binge Drinking

By Andrew M. Seaman

Young people who are more receptive to alcohol ads on TV may be at higher risk of problem drinking over the next few years, according to a new study.

"If you compare low- to high-receptivity kids, their risk of transitioning to binge drinking was over four times higher," said Dr James Sargent, the study's senior author from the Geisel School of Medicine at Dartmouth College in Lebanon, New Hampshire.

Sargent and his colleagues write in *JAMA Pediatrics* that in 2013, about two thirds of US high school students reported drinking. About a third reported drinking in the past month, and about one in five reported recent binge drinking, that is, five or more drinks on one occasion.

Previous research tried to establish a link between TV alcohol advertisements and young people's drinking behaviours, but with conflicting results.

For the new study, the researchers applied a method previously used to find a link between smoking shown in movies and people's smoking behaviour. The method involves showing people ads stripped of brands, to see what they can recall from having seen the ad on TV.

In 2010 and 2011, more than 3,000 people ages 15 to 23 answered a series of questions over the phone and then finished the image portion of the study online. Two years later, 1,596 participants completed follow-up surveys.

The youngest participants were only slightly less likely than the oldest ones - about 23 per cent versus 26 per cent - to report having seen alcohol ads, to like the ads they saw and to identify the alcohol brands in the ads.

Liking and remembering ads was considered a sign of greater receptivity to the advertising message. And at the two-year follow-up, participants who had scored highest for receptivity were more likely to have transitioned to drinking, binge drinking and hazardous drinking.

"This study suggests that alcohol marketing does affect subsequent drinking behaviours," Sargent said.

He said they also checked for a link between fast food TV ads and drinking behaviours. There was no connection, which reduces the chance that children who progressed to drinking were simply more susceptible to ads in general.

Sam Zakhari, chief scientist of the Distilled Spirits Council, wrote in a statement to Reuters Health that new research is driven by advocacy – not science.

"The clearest indication of this is that according to the US Government, underage drinking is at historic lows, yet advertising and marketing are at all-time highs," Zakhari said. "The multiple flaws in this study undercuts the credibility of its conclusions."

Zakhari, a former director at the National Institute on Alcohol Abuse and Alcoholism, said, "Research shows that advertising does not cause someone to begin drinking alcohol or to drink more."

Previous studies have found links between advertising and drinking behaviours, however.

Last year, researchers found that the preferred brands of alcohol among underage drinkers match brands advertised in the most popular magazines in that age group (see Reuters Health story of 8 July 8 2014 here: http://reut.rs/1EsFPOc).

David Jernigan, lead author of the 2014 study, said the new work by Sargent and colleagues is "yet another study showing that exposure to alcohol advertising on TV is associated with young people progressing to more hazardous drinking".

"This is the kind of research we need to inform a robust policy debate about what we can do that will actually protect kids," said Jernigan, who directs the Center on Alcohol Marketing and Youth at the Johns Hopkins Bloomberg School of Public Health in Baltimore and was not involved in the new study.

Jernigan said his research suggests there are ways for companies to target people of legal drinking age on TV without influencing underage viewers. One way is to target ads at people closer to 30 and older.

"Alcohol advertising is aimed at 21- to 25- or 28-year olds," said Sargent. "Given the similarity in terms of psychology of 21-year-olds and 18-year-olds and 17-year-olds, and the similarity of the programmes that they watch, it's really absurd that you can have advertisements that target 21-year-olds without influencing a 17- or 18-year-old. It's just really common sense."

24 March 2015

⇨ The above information is reprinted with kind permission from *Huffington Post*. Please visit www.huffingtonpost.co.uk for further information.

© 2018 AOL (UK) Limited

Alcohol adverts seen "almost once a minute" during Euro 2016 games

TV viewers frequently saw slogan of tournament sponsor Carlsberg throughout England and Wales matches, says charity.

Football supporters watching the England and Wales matches during the group stages of Euro 2016 saw alcohol marketing almost once a minute during game play, a charity has said.

With French laws banning alcohol sponsorship of sporting events and alcohol advertising on television, the Euro 2016 sponsor Carlsberg replaced its brand name on pitch-side digital boards with one of its well-known slogans, in the brand's font, Alcohol Concern said.

Over the five group matches played by England and Wales – including the game between them that ended in a 2-1 victory for England – these slogans were seen 392 times, the charity said.

This equated to an average of 78.4 a game, or once every 72 seconds.

Tom Smith, the director of campaigns at Alcohol Concern, said: "The volume of alcohol marketing in sport, especially in football, which is popular with children and younger people, is enormous. We already know from our previous research that half of children associate leading beers with football.

"Alcohol marketing drives consumption, particularly in under-18s, and sport should be something which inspires active participation and good health, not more drinking."

He said the Government should phase out alcohol marketing from sport, as it has done with tobacco.

A Carlsberg spokesman said: "We take great care that the vast majority of viewers of our marketing are above legal drinking age. Our internal and industry codes clearly stipulate that our marketing communications are designed to prevent any primarily underage appeal."

27 June 2016

⇨ The above information is reprinted with kind permission from The Press Association. Please visit www.theguardian.com for further information.

The impact of alcohol advertising

An extract from the ELSA project report on the evidence to strengthen regulation to protect young people.

Young people and commercial communications

The adolescent brain undergoes major development, which makes them more vulnerable to impulsivity and greater sensitivity to pleasure and reward. Young people who already have problems related to alcohol are likely to be particularly vulnerable to alcohol advertising, with the vulnerability increasing with increasing alcohol consumption. Alcohol advertising manipulates adolescents' vulnerability by shaping their attitudes, perceptions and particularly expectancies about alcohol use, which then influence youth decisions to drink. All types of media are used for commercial communications, including television, music and music videos, films, paid placements in films and TV shows, the Internet, grass roots word of mouth, and sports sponsorship. Some of these media are easy to regulate (for example television and sports sponsorship), whereas others are more difficult to regulate, including the Internet, which appears to be increasing as a medium for alcohol commercial communications. There is considerable evidence, although mostly from the United States, that commercial communications for alcohol are targeted to young people.

Are young people vulnerable to alcohol advertising?

Adolescents have three distinctive vulnerabilities: impulsivity, linked to a temporal gap between the onset of hormonal and environmental stimuli into the amygdala and the more gradual development of inhibitory control through the executive planning and decision-making functions of the prefrontal cortex; self-consciousness and self-doubt, attributable at least in part to the emergence of abstract thinking, but evident in the greater frequency and intensity of negative mood states during adolescence; and elevated risk from product use, including impulsive behaviour such as drinking and driving, but also greater susceptibility to toxins because of the plasticity of the developing brain as well as greater sensitivity to the brain's 'stamping' functions identifying pleasure and reward.

Adolescents aged 14 to 17 years with alcohol use disorders showed substantially greater brain activation to alcoholic beverage pictures than control youths, predominantly in brain areas linked to reward, desire, and positive effect. The degree of brain response to the alcohol pictures was highest in youths who consumed more drinks per month and reported greater desires to drink.

Early work on alcohol advertising and youth tended to rest on a simple theoretical basis: exposure to alcohol advertising influences youth drinking behaviour.

However, more recent studies have pointed to the importance of alcohol advertising in shaping youth attitudes, perceptions and particularly expectancies about alcohol use, which then influence youth decisions to drink. There seems to be a cognitive progression from liking of alcohol advertisements (an affective response associated with the desirability of portrayals in the advertisements and a resulting identification with characters in the advertisements) to positive expectancies about alcohol use, to intentions to drink or actual drinking among young people.

What young people appear to like in alcohol advertising is elements of humour and story, with somewhat less appreciation of music, animal characters and people characters. Liking of these elements significantly contributes to the overall likeability of specific advertisements, and then to greater likelihood of intent to purchase the brand and product advertised.

Cross-sectional analysis of an American study found that adolescents progressively internalised messages about alcohol, and that these messages affected their drinking behaviours. Young people who watched more primetime television found portrayals of alcohol in alcohol advertising more desirable, and showed greater desire to emulate the persons in the advertisements. These were associated with more positive expectancies about alcohol use, which then positively predicted liking beer brands as well as alcohol use. Further analysis found a positive relationship between liking of alcohol advertisements at baseline and alcohol consumption over a follow-up period of three years, among a cohort of nine- to 16-year-olds from nine counties in the San Francisco Bay Area. The effects of liking the advertisements were mediated through expectancies about alcohol use, as well as through normative effects of the exposure to alcohol advertising. Young people who liked alcohol advertising not only believed that positive consequences of drinking were more likely, but also were more likely to believe that their peers drank more frequently, and that their peers approved more of drinking. All these beliefs interacted to produce a greater likelihood of drinking, or of intention to drink within the next year.

What types of media are used for commercial communications?

Television portrayal of alcohol use has been given a lot of attention. When people are seen drinking on television they seem to be drinking alcohol most of the time, for example, found that every 6.5 min a reference to alcohol was made in their sample of 50 programmes on British television. Especially in fictional series the consumption of alcohol was prominently present. concentrated on the portrayal of alcohol and drinking in six British soap operas and concluded that 86% of all programmes contained visual or verbal references to alcoholic beverages.

More alcohol was consumed than any other kind of drink, but the sample of programmes almost never referred to the hazards of alcohol consumption.

Content analyses of portrayals of alcohol use on television suggest that incidences of drinking occur frequently and that these portrayals present drinking as a relatively consequence-free activity. Television characters who drink tend to be 'high status' characters who are wealthy, successful, attractive, and in senior-level occupations. Their drinking is often associated with happiness, social achievement, relaxation and camaraderie.

Content analyses of the appeals used in alcohol advertisements suggest that drinking is portrayed as being an important part of sociability, physical attractiveness, masculinity, romance, relaxation and adventure. Many alcohol advertisements use rock music, animation, image appeals and celebrity endorsers, which increases their popularity with underage television viewers. Not surprisingly, then, alcohol commercials are among the most likely to be remembered by teenagers and the most frequently mentioned as their favourites.

Music and music videos

An analysis of music that is popular with youth found that 17% of lyrics across all of the genres contained references to alcohol. Alcohol was mentioned more frequently in rap music (47%) than in other genres, such as country-western (13%), top 40 (12%), alternative rock (10%) and heavy metal (3%). A common theme is getting intoxicated or high, although

drinking also is associated with wealth and luxury, sexual activity, and crime or violence. As with television and film, consequences of drinking are mentioned in few songs and anti-use messages occur rarely. Product placements or brand name mentions occurred in approximately 30% of songs with alcohol mentions and are especially common in rap music (48%). From 1979 to 1997, rap music song lyrics with references to alcohol increased fivefold (from 8% to 44%); those exhibiting positive attitudes rose from 43% to 73%; and brand name mentions increased from 46% to 71%. There were also significant increases in songs mentioning champagne and liquor (mainly expensive brand names) when comparing songs released after 1994 with those from previous years. In addition, there were significant increases in references to alcohol to signify glamour and wealth, and using alcohol with drugs and for recreational purposes. The findings also showed that alcohol use in rap music was much more likely to result in positive than negative consequences.

A similar pattern is found for music videos. found that rap music videos contained the highest percentage of depictions of alcohol use, whereas rhythm and blues videos showed the least alcohol use. Additionally, alcohol use was found in a higher proportion of music videos that had any sexual content than in videos that had no sexual content. Both the content, which has been shown to glamorizs the use of alcohol, and the advertisements surrounding the music videos have a potential to make drinking alcohol more enticing to young viewers.

Internet

The rapid rise of information technology and, in particular, the Internet has given manufacturers a new promotional opportunity. Sophisticated web sites have been created using technology to produce interactive arenas with impressive graphics and eye-catching animation. Research on alcohol portrayals on the Internet has focused on youth access, exposure to alcohol marketing, and the potential attractiveness of commercial alcohol web sites to youth. Research

has not addressed the content of non-commercial web sites that focus on alcohol products or drinking cultures. Similarly, no study has addressed the potential effects on consumption by youth of exposure to alcohol portrayals and promotion on the Internet. The Center for Media Education found that commercial alcohol web sites are easily accessible to youth, and are often accessed from search engines through non-related key word searches for games, entertainment, music, contests and free screensavers. Content analyses of web sites that are registered to large alcohol companies revealed that young drinkers are targeted through a glorification of youth culture that offers humour, hip language, interactive games and contests, audio downloads of rock music, and community-building chat rooms and message boards. Overall, these sites were found to promote alcohol use. Only a handful of them included any information on the harm done by alcohol.

Are commercial communications targeted to young people?

Research in the United States shows that alcohol companies have placed significant amounts of advertising where youth are more likely per capita to be exposed to it than adults. In 2002 in the US, underage youth saw 45% more beer and ale advertising, 12% more distilled spirits advertising, 65% more low-alcohol refresher advertising, and 69% less advertising for wine than persons 21 years and older. Girls aged 12 to 20 years were more likely to be exposed to beer, ale and low-alcohol refresher advertising than women in the group aged 21 to 34. Girls' exposure to low-alcohol refresher advertising increased by 216% from 2001 to 2002, while boys' exposure increased by 46%.

Magazines are the most tightly targeted of the measured media. Two studies to date have looked at alcohol advertising in this medium. Following on research suggesting that cigarette brands popular among youth ages 12 to 17 were more likely than other brands to be advertised in magazines selected a convenience sample of 15 magazines, 11 with the

highest youth readership (greater than 1.9 million readers) and four with the lowest youth readership (less than 0.8 million), and assessed the volume of influence by counting advertising pages for alcohol and tobacco in each magazine. The authors found a relationship between the size of youth readership and alcohol and tobacco advertisements, with magazines with more youth readers containing more alcohol and tobacco advertisements.

Conclusions

Due to the biological changes of adolescence, young people are particularly vulnerable to alcohol advertisements, which manipulate adolescents' vulnerability by shaping their attitudes, perceptions and expectancies about alcohol use.

This vulnerability is exacerbated by the enormous exposure to commercial communications, not only through traditional media, which are highly targeted to young people, but also through mobile phones and the Internet, which have particular appeal to young people. Article 95(3) of the Treaty of the European Union requires the Commission, in its proposals for the establishment and functioning of the Internal Market concerning health, to take as a base a high level of protection. An approximation of the European countries' advertising laws, including statutory regulations and a ban in certain media would protect young people, by regulating the promotion of alcohol, an addictive product responsible for over 25% of young male deaths and over 11% of young female deaths, and avoid a situation where young people begin using alcohol at an early age and in a risky fashion as a result of promotion and thereby become dependent.

⇨ The above extract is reprinted with kind permission from ELSA. Please visit www.europa.eu for further information.

© 2018 European Health Project

Captain Morgan rum ad banned for implying alcohol increases confidence

TV advert showing party on a ship attracted complaints for implying a social occasion's success depends on alcohol.

By Josie Clarke

An ad for a brand of rum sold by drinks giant Diageo has been banned for implying that alcohol can increase confidence.

The television ad for Captain Morgan showed a party on a sailing ship and a man, with the face of the Captain Morgan character superimposed over his own, dancing, upending a sofa and swinging on a rope between decks.

Alcohol Concern and a member of the public complained that the ad suggested that alcohol could contribute to a drinker's popularity or confidence and that the success of a social occasion depended on alcohol.

Diageo said the ad emphasised the attitude that the brand "embodied", of camaraderie, enjoying time with friends and living life to the full.

The company said no alcohol was shown in the party scenes, and they did not think that there was anything to suggest that the individuals shown had consumed or would consume alcohol.

The Advertising Standards Authority (ASA) said viewers were likely to understand that the central figure's behaviour resulted from his consumption of Captain Morgan rum.

The ASA said: "Although the ad did not explicitly depict drinking alcohol as resulting in a change in the central character's behaviour in a 'before and after' scenario, we considered that the superimposed Captain Morgan face implied that he had already consumed the product and thus linked his confident behaviour to this consumption.

"We concluded that the ad implied that drinking alcohol could enhance personal qualities and was therefore irresponsible."

However, the ASA said the ad did not imply that the general success of the party was dependent on the presence or consumption of alcohol.

It ruled that the ad must not appear again in its current form and told Diageo not to imply that alcohol could enhance confidence.

Julie Bramham, European marketing director for Captain Morgan, said: "Whilst we are pleased that the ASA chose to not uphold part of the complaint, we disagree with their interpretation on the rest of the ruling.

"No alcohol was pictured and the Captain Morgan face was designed to represent the brand as a whole and not intended to be linked to the consumption of alcohol."

31 August 2016

⇨ The above information is reprinted with kind permission from The Press Association. Please visit www.independent.co.uk for further information.

Alcohol marketing and children

Much of the debate around alcohol advertising concerns the possible effects on children and young people. The Advertising Codes prohibit the specific targeting of minors, but the ubiquity of alcohol advertising ensures that they can hardly miss it. Ofcom report that since a period of gradual decline of exposure to television alcohol advertising for children aged 10–15 from 2002 to 2006, 2007 to 2011 represented a period of absolute increase in exposure.[1] With the proliferation of online streaming services, there has also been shown to be a "substantial potential" for young people to be exposed to alcohol advertising through Internet television.[2] Research examining exposure to television alcohol advertising indicated that 10- to 15-year-olds in the UK were significantly more exposed to alcohol advertisements per viewing hour than adults (25 years and over).[3]

In the UK, the proportion of children drinking alcohol remains well above the European average.[4] Evidence shows that exposure to alcohol marketing encourages children to drink at an earlier age and in greater quantities than they otherwise would. The Science Committee of the European Alcohol and Health Forum concluded in 2009 that "alcohol marketing increases the likelihood that adolescents will start to use alcohol, and to drink more if they are already using alcohol".[5] Longitudinal research from Europe has suggested that adolescents' alcohol use is affected by exposure to alcohol marketing,[6] while research from the US has suggested that exposure to alcohol-related media may in fact begin a mutually influencing process in adolescents, escalating alcohol use over time; the more alcohol-related media they see, the more they drink, and the more they drink, the more they seek out alcohol-related media.[7]

Indeed, the evidence is that even young children are aware of alcohol advertisements and tend to remember them. Manufacturers further reduce the chances of young people failing to get the message by sponsorship of sports teams and events and music concerts having particular appeal to the young. A 2016 systematic review of seven studies exploring alcohol sports sponsorship found a positive association between exposure to such marketing and alcohol consumption, with two of the studies reviewed showing this relationship held for schoolchildren.[8] There is also evidence that underage drinking and the likelihood of alcohol problems in later life are closely related to positive expectations of benefits from alcohol use, precisely the expectancies advertising is designed to encourage.[9]

American studies have found that children and teenagers respond particularly positively to TV advertisements featuring animals, humour, music and celebrities. It is suggested, therefore, that policy makers should ensure that advertisements should focus on product-related characteristics, using content less appealing to children and teenagers.[10]

An American study found that heavy advertising by the alcohol industry in the US has such considerable influence on adolescents that its removal would lower underage drinking in general and binge drinking in particular. The analysis suggested that the complete elimination of alcohol advertising could reduce monthly drinking by adolescents from about 25% to about 21%, and binge drinking from 12% to around 7%. However, these estimated reductions were substantially less than those which the analysis suggested would result from significantly increasing the price of alcoholic drinks.[11]

Another American study found that youth who saw more alcohol advertisements drank more on average, each additional advertisement seen increasing the number of drinks consumed by 1%. Also, youth in markets with greater alcohol advertising expenditures drank more, each additional dollar spent per capita increasing the number of drinks consumed by 3%. Youth in markets with more alcohol advertisements showed increased drinking levels into their late 20s whereas drinking plateaued in the early 20s for youth in markets with fewer advertisements.[12]

A study of the impact of alcohol advertising on teenagers in Ireland found:[13]

⇨ Alcohol advertisements were identified as their favourite type of advert by the majority of those surveyed

⇨ Most of the teenagers believed that the majority of the alcohol advertisements were targeted at young people. This was because the advertisements depicted scenes – dancing, clubbing, lively music, wild activities – that identified with young people

⇨ The teenagers interpreted alcohol advertisements as suggesting, contrary to the code governing alcohol advertising, that alcohol is a gateway to social and sexual success and as having mood altering and therapeutic properties

A review of seven international research studies[14] concluded that there is evidence for an association between prior alcohol advertising and marketing exposure and subsequent alcohol drinking behaviour in young people. The forms of exposure included both direct exposure to advertising using broadcast and print media, and indirect methods such as in-store promotions and portrayal of alcohol drinking in films, music videos and TV programmes. Three studies showed that onset of drinking in adolescent non-drinkers at baseline were significantly associated with exposure to alcohol marketing. One study showed that for each additional hour of TV viewing per day, the risk of starting to drink increased by 9% during the following 18 months. Another found that youth with higher exposure to alcohol use depicted in popular movies were more likely to have tried alcohol

13 to 26 months later. Yet another showed that exposure to in-store beer displays significantly predicted drinking onset two years later. Two studies demonstrated dose response relationships. In one, in Flemish school children, increased frequency of TV viewing and music video viewing was highly significantly related to the amount of alcohol consumed while going out. In the other, of individuals aged 15 to 26 years, for each additional advertisement seen the number of drinks consumed increased by 1%, and for each additional dollar spent per capita on alcohol advertisements the number of drinks consumed increased by 3%.

A US study further found that receptivity to alcohol advertising on television predicted the onset of drinking, binge drinking and hazardous drinking for young people aged 15 to 23 years.[15] A similar European study found that for adolescents, naming a favourite alcohol advertisement increased their likelihood of beginning to binge drink within the next year.[16] Recent findings from the UK found that exposure to alcohol use in films was linked with higher risk of alcohol use and alcohol-related problems in adolescents.[17] Such a link between alcohol use in movies and adolescent binge drinking was found to be relatively stable across cultures in a 2012 European study.[18]

Social media and online represent a substantial marketing channel for many brands; in April 2016, Facebook was visited by 38.9 million unique users in the UK alone, whilst Twitter and Instagram received 20.9 million and 16.5 million, respectively.[19] Australian researchers investigating drinking behaviours of 15-to 29-year-olds have found an association between liking or following alcohol social media profiles and riskier alcohol consumption.[20] US research found YouTube profiles created for fictional users aged 14, 17 and 19 were able to subscribe to 100% of the alcohol brand YouTube pages explored.[21] Research from Australia found the alcohol websites they investigated typically had poor filter systems protecting underage visitors.[22]

References

[1] Ofcom (May 2013) 'Children's and young people's exposure to alcohol advertising 2007 to 2011', p. 7 <https://www.ofcom.org.uk/__data/assets/pdf_file/0018/51507/alcohol_report_2013.pdf>

[2] Siegel, M., Kurland, R., Castrini, M., Morse, C., de Groot, A., Retamozo, C., Roberts, S., Ross, C., and Jernigan, D. (2016) 'Potential youth exposure to alcohol advertising on the internet: a study of internet versions of popular television programs', Journal of Substance Use, 21: 4, pp. 361–367

[3] Patil, S., Winpenny, E., Elliott, M., Rohr, C., and Nolte, E. (2014) 'Youth exposure to alcohol advertising on television in the UK, the Netherlands and Germany', The European Journal of Public Health, 24: 4, pp. 561–565 <http://eurpub.oxfordjournals.org/content/24/4/561>

[4] Public Health England (July 2016) 'Data intelligence summary: Alcohol consumption and harm among under 18-year-olds' p. 14

[5] Scientific Opinion of the Science Group of the European Alcohol and Health Forum (2009) 'Does marketing communication impact on the volume and patterns of consumption of alcoholic beverages, especially by young people? – a review of longitudinal studies'

[6] Bruijn, A., Tanghe, J., Leeuw, R., Engels, R., Anderson, P., Beccaria, F., Bujalski, M., Celata, C., Gosselt, J., Schreckenberg, D. and Słodownik, L. (2016) 'European longitudinal study on the relationship between adolescents' alcohol marketing exposure and alcohol use.' Addiction, 111: 10, pp. 1774–1783.

[7] Tucker, J., Miles, J. and D'Amico, E. (2013) 'Cross-lagged associations between substance use-related media exposure and alcohol use during middle school.' Journal of Adolescent Health, 53: 4, pp. 460–464.

[8] Brown, K. (February 2016) 'Association Between Alcohol Sports Sponsorship and Consumption: A Systematic Review', Alcohol and Alcoholism, 2016, pp. 1–9

[9] Hill, L., and Casswell (2001) 'Alcohol Advertising and Sponsorship: Commercial Freedom and Control in the Public Interest' in Heather, N., Peters, J. S., and Stockwell, T (eds)., 'International Handbook of Alcohol Dependence & Problems', John Wiley & Sons

[10] Chen, M. et al (September 2006) 'Alcohol advertising: What makes it attractive to youth?', Journal of Health Communications, 10, pp. 553–565, Routledge Taylor & Francis Group <http://www.ncbi.nlm.nih.gov/pubmed/16203633#_blank>

[11] Saffer, H., and Dave, D. (May 2003). 'Alcohol Advertising and Alcohol Consumption by Adolescents', NBER Working Paper No. 9676

[12] Snyder, L. B, Milici, F., Slater, M., Sun, H., Strizhakova, Y. (January 2006). 'Effects of Alcohol Advertising Exposure on Drinking Among Youth', Archives of Pediatrics & Adolescent Medicine, 160: 1, pp. 18–24 <http://archpedi.jamanetwork.com/article.aspx?articleid=204410#_blank>

[13] Dring, C., Hope, A. (November 2001). 'The Impact of Alcohol Advertising on Teenagers in Ireland', Health Promotion Unit, Department of Health & Children

[14] Smith, L., and Foxcroft, D. (November 2007). 'The effect of alcohol advertising and marketing on drinking behaviour in young people: systematic review of published longitudinal studies', Alcohol Education and Research Council, now Alcohol Research UK

[15] Tanski, S., McClure, A., Li, Z., Jackson, K., Morgenstern, M., Li, Z. and Sargent, J. (2015) 'Cued recall of alcohol advertising on television and underage drinking behavior.' JAMA Pediatrics, 169: 3, pp. 264–271.

[16] Morgenstern, M., Sargent, J., Sweeting, H., Faggiano, F., Mathis, F. and Hanewinkel, R. (2014) 'Favourite alcohol advertisements and binge drinking among adolescents: a cross cultural cohort study.' Addiction, 109: 12, pp. 2005–2015.

[17] Waylen, A., Leary, S., Ness, A., and Sargent, J. (May 2015) 'Research has also indicated that exposure to alcohol use in films is associated with higher risk of alcohol use and alcohol-related problems in adolescents in the UK.' Paediatrics, 135: 5, pp. 851–858 <http://pediatrics.aappublications.org/content/135/5/851>

[18] Hanewinkel, R., Sargent, J., Poelen, E., Scholte, R., Florek, E., Sweeting, H., Hunt, K., Karlsdottir, S., Jonsson, S., Mathis, F. and Faggiano, F. (2012) 'Alcohol consumption in movies and adolescent binge drinking in 6 European countries.' Pediatrics, 129: 4, pp. 1–12.

[19] Ofcom (August 2016) 'The Communications Market 2016', p. 181 <https://www.ofcom.org.uk/__data/assets/pdf_file/0024/26826/cmr_uk_2016.pdf>

[20] Carrotte, E., Dietze, P., Wright, C. and Lim, M. (2016) 'Who 'likes' alcohol? Young Australians' engagement with alcohol marketing via social media and related alcohol consumption patterns.' Australian and New Zealand Journal of Public Health, 40: 474–479 <http://onlinelibrary.wiley.com/doi/10.1111/1753-6405.12572/abstract>

[21] Barry, A., Johnson, E., Rabre, A., Darville, G., Donovan, K., and Efunbumi, O. (2014) 'Underage Access to Online Alcohol Marketing Content: A YouTube Case Study', Alcohol and Alcoholism, 50: 1, pp. 89–94 <http://dx.doi.org/10.1093/alcalc/agu078>

[22] Jones, S., Thom, J., Davoren, S. et al. (2014) 'Internet filters and entry pages do not protect children from online alcohol marketing' Journal of Public Health Policy, (2014), 35: 1, pp. 75–90 <http://doi:10.1057/jphp.2013.46>

7 July 2017

⇨ The above information is reprinted with kind permission from the Institute of Alcohol Studies. Please visit www.ias.org.uk for further information.

Public health experts call for ban on alcohol advertising in UK

Global study claims industry's marketing methods are encouraging young people to drink and often breach codes of practice.

By Damien Gayle

Public health experts have called for a ban on alcohol advertising in the UK in light of new research that claims the industry's marketing practices encourage young people to drink.

Studies into the impact of alcohol advertising around the world found that marketing practices often seemed to breach the industry's own voluntary codes of practice, which in any case were not sufficient to protect children.

"It is clear that self-regulation is not working and we welcome calls for greater action from governments to protect children from exposure to alcohol marketing," said Professor Sir Ian Gilmore, chair of the Alcohol Health Alliance (AHA), an umbrella group of more than 40 UK health NGOs, including the Academy of Medical Royal Colleges.

The call comes despite recent data showing that levels of youth drinking in the UK are the lowest on record, with only about 17% of children aged eight to 15 admitting to ever drinking alcohol.

> "We know that alcohol marketing contains content and messages that appeal to children, and that due to exposure to this advertising, children drink more, and start drinking at an earlier age"

AB Inbev and Diageo, two of the world's biggest alcoholic drinks makers, have reported ploughing as much as 15% of their annual global sales back into marketing, amounting to $7 billion (£5.75 billion) and £1.6 billion respectively. The new research, in the journal *Addiction*, investigated the techniques of alcohol marketing and their effects on young people.

The findings of the 14 studies included claims that marketing practices around the 2014 FIFA World Cup appeared to breach voluntary codes of practice, that young people were increasingly exposed to alcohol marketing through social media, and that much of the marketing included content likely to appeal to young people.

The lead editor, Professor Thomas Babor from the University of Connecticut, said: "No other legal product with such potential for harm is as widely promoted and advertised in the world as alcohol. These papers provide a wealth of information to support governments in their efforts to protect children and other vulnerable populations from exposure to alcohol marketing."

In the UK, advertising for alcoholic drinks follows a code enforced by the Advertising Standards Authority, while the packaging and branding of the products is subject to self-regulation. One recent drinks advert judged to be in breach of standards was a promotional film for Captain Morgan rum featuring a raucous party aboard a sailing ship, which was banned for implying alcohol can make you more confident.

Ian Hamilton, a specialist in problem drug use at the University of York, who was not involved in any of the *Addiction* papers, said alcohol marketing had found ways to evade statutory limits, particularly on social media and through endorsements of music and sporting events.

"Some of the messages are quite subtle, but they are persistent," he said. "So this idea that alcohol is necessary for social success, or is both a stimulant as well as a sedative, that it removes sexual inhibition, that it improves – bizarrely – your sporting and mental abilities.

"Of course, the way they do it is they don't say go and buy Carlsberg, but they'll do endorsed interviews with celebrities or they'll offer free music downloads or notices of events, so they do it in quite subtle and clever ways."

A spokesperson for the AHA said while a comprehensive ban – similar to one in place in Norway – was the ultimate aim, in the meantime the organisation would like to see an interim solution that began with stopping sports sponsorship, introducing a watershed for alcohol adverts on television, and restricting cinema adverts so that they could be shown only before 18 certificate films.

Such a partial ban would still allow adverts in newspapers, billboards and radio, but would restrict them to providing basic information including a product's origin, strength and production method, the spokesman suggested. Such a framework would be far stronger than the current ASA code of practice, which only stipulates what cannot be included in an advert.

"Tighter alcohol marketing regulation in the UK, without industry involvement, is desirable, achievable and effective," said Paul Lincoln, the chief executive of UK Health Forum, which is a member of the AHA.

The claims and recommendations were disputed by the alcohol industry. Dave Roberts, the director general of the Alcohol Information Partnership, an industry-funded group, said levels of underage and harmful drinking had been falling year on year in the UK.

He said: "Official data shows the vast majority of people drink in moderation and in a convivial manner.

A self-regulatory framework and a partnership approach have clearly been working.

"The best way to reduce alcohol-related harm is to target programmes and policies at harmful drinkers. Instead of restricting companies' freedom to operate and compete it would be better for government to focus on understanding what has worked so well over the past decade and encourage more of the same. Where there are pockets of harm, intervention should be directed towards those communities or age groups."

"Hamilton also warned that a blanket ban on alcohol advertising could be seen as 'taking a sledgehammer to crack a nut', with the danger that such a policy could give drinking a kind of outlaw prestige that might increase its appeal to some"

But, he added, a similar policy on tobacco advertising had apparently proved successful in diminishing the appeal of smoking. "I think the state does have some kind of responsibility," Hamilton said. "We can't have do-it-yourself regulation by industry whose prime motive is to find the next generation of consumers."

10 January 2017

⇨ The above information is reprinted with kind permission from *The Guardian*. Please visit www.theguardian.com for further information.

Cutting alcohol ads in sport sends the right message to youngsters THE CONVERSATION

*An article from **The Conversation.***

By Susan Goldstein, Honorary Senior Lecturer at the School of Public Health, University of the Witwatersrand

An English barbershop owner and his son have embarked on a massive David and Goliath battle to outbid Thai beer Chang as the front-of-jersey sponsor of their local football team.

In what they call the #OutbidChang campaign they started a crowd fund to stop Everton Football Club in the UK from using the beer as its sponsor. Everton FC is the last soccer team in the English Premier League to be funded by an alcohol company.

The father and son duo argues that alcohol sponsorship fuels alcohol harm to children and that in the UK children are more familiar with alcohol brands than with brands of biscuits.

The campaign is being supported by BigAlcohol.Exposed – a global network of non-governmental organisations dedicated to exposing the truth about the unethical business methods of the alcohol industry.

Whether the duo is indeed able to outbid Chang beer remains to be seen – but the cause is a valid one.

The reality is that in 2012, across the globe about 3.3 million deaths – or 5.9% of all deaths – were attributable to alcohol consumption. Alcohol ranks among the top five risk factors for disease, disability and death throughout the world, according to the World Health Organization.

It results in more deaths than HIV/AIDS and TB, and is the causal factor in more than 60 major types of disease, including neuropsychiatric disorders like epilepsy, gastro-intestinal disorders such as liver cirrhosis, cancer, cardio-vascular diseases and diabetes.

There is no doubt that the abuse of alcohol is harmful. And it is clear that

the harm is greater in countries where there is "harmful" or binge drinking.

South Africa is one of these countries. Despite the fact that 65% of adults do not drink, those who do, drink 35 litres of absolute alcohol a year. This is among the highest in the world.

The expressed aim of any alcohol-producing company is to make profit – and to do this the more that is sold, the better. In South Africa the spend on above-the-line advertising in 2012 was R1.8 billion (about US$114 million at current rates). And that excluded sponsorship and other marketing opportunities.

In the US, alcohol advertisers spent $2 billion on alcohol advertising in measured media – television, radio, print, outdoor, major newspapers and Sunday supplements – in 2005.

Because the adverts were placed in media the youth would likely engage with, US youth viewed 45% more beer ads and 27% more liquor ads in magazines than people of legal drinking age. They also watched an average of 2,000 television ads for alcohol per year.

There is extensive research showing that young people who are exposed to alcohol advertising are more likely to intend to drink, start drinking at an early age and to drink more than those not exposed.

Does banning advertising work?

Unlike many harmful substances, alcohol is legal in most societies, and is freely advertised and promoted. This includes marketing and promoting excessive use, as seen through the creation of the six pack and 12 pack, promoting high rates of use.

But there is evidence that banning the advertising of a product does produce results.

Take tobacco advertising as an example. Cigarettes are more addictive than alcohol. And each year about six million people die prematurely from tobacco-related illnesses.

A recent World Health Organization report found that in 2010 – seven years after the organisation introduced a tobacco monitoring framework that suggested advertising bans – there were 3.9 billion non-smokers aged 15 years and over in the organisation's member states, or 78% of the 5.1 billion population aged over 15. This number is projected to rise to five billion – or 81% of the projected 6.1 billion population aged over 15 – by 2025 if the current pace of tobacco cessation continues.

And an analysis of advertising spend in 2001, when the tobacco advertising ban came into effect in South Africa, shows that economic growth was only 2.7% and advertising decreased by 7.4%. But by 2002 the adspend was back to the 2001 level as the industry found new ways to promote its product. It continued to grow until 2007 when it was approximately double the amount spent in 2002.

The lesson to learn with banning tobacco advertising is that there will always be products to jump into the breach and the industry will continue to grow.

Sending the wrong message

Many argue that, as alcohol is a legal substance, the industry must have the right to advertise. In recognition of the harm that the industry admits that alcohol causes, a voluntary code has been developed that the industry is supposed to adhere to.

This code is published on the web page of the Advertising Standards Authority of South Africa. One of the guidelines is that commercial communication may not imply that alcohol consumption is essential to business and/or social success or acceptance, or that refusal to consume is a sign of weakness.

What isn't clear is how this guidance is interpreted, monitored or enforced.

As a result, for example, a leading beer brand in South Africa sponsors several events – both in sport and

entertainment – as well as many male-dominated cultural activities. One of its competitions – linked to a football event – asked participants to enter by buying a "pack" of beer.

Football is a major preoccupation of young boys. And although the industry argues that is it advertising to adults, all boys are encouraged to participate in and watch these sports. Information about these events is freely available to boys and girls under 18.

A study in the US has found that while 26% of young adults between the ages of 21 and 23 had seen an alcohol advertisement, 23% of 15- to 17-year-olds had also seen the same advert. They also found that young people who could accurately identify alcoholic products and who said they liked the ads were more likely to try drinking or to drink more.

In South Africa all three predominantly male national sports – football, rugby and cricket – are sponsored by alcohol.

Although the South African Government in May 2015 gazetted a new National Liquor Policy for comment recommending that alcohol advertising be restricted, and sponsorship and promotions associated with alcohol be prohibited, whether this will make it into the final legislation remains to be seen.

Considering South Africa's high levels of alcohol abuse and harm, this would be a step in the right direction. Whether ordinary South Africans would take a similar stand to the English barbershop owner and his son and crowd fund an outbid campaign against liquor giants is another question.

31 May 2016

⇨ The above information is reprinted with kind permission from *The Conversation*. Please visit www.theconversation.com for further information.

Key facts

- Alcohol "a direct cause of seven types of cancer". (page 1)

- Alcohol is a well-known disinfectant and some have speculated it may be useful for treating gut infections. (page 4)

- 14 units is equivalent to six pints of average-strength beer or ten small glasses of low-strength wine. (page 6)

- Regularly drinking more than 14 units a week risks damaging your health. (page 6)

- In the most conservative countries, "low-risk" consumption means drinking no more than 10g of alcohol per day for women and 20g for men. (page 8)

- But in Chile, a person can down 56g of alcohol per day and still be considered a low-risk drinker. (page 8)

- A unit translates to 10ml, or 8g, of pure alcohol – the amount of alcohol the average adult can process in an hour. (page 8)

- A one-unit alcoholic drink is roughly equivalent to 250ml of 4% strength beer, 76ml of 13% wine or 25ml of 40% spirits. (page 8)

- Alcohol-specific deaths fell to 17,755 in the three year period 2012 to 2014. Down 3% compared to the previous three-year period (3.1% fall among men and a 2.5% fall among women). (page 9)

- If you have just discovered you are pregnant and you have been drinking then you shouldn't automatically panic as it is unlikely in most cases that your baby has been affected; though it is important to avoid further drinking. (page 10)

- One in five girls (and one in ten boys) aged 14 to 15 goes further than they wanted to in a sexual experience after drinking alcohol. In the most serious cases, alcohol could lead to them becoming the victim of a sexual assault. (page 11)

- If a child or young person drinks alcohol, then they are more likely to get into trouble with the police. Every year in the UK, more than 10,000 fines for being drunk and disorderly are issued to young people aged 16 to 19. (page 11)

- Drinking alcohol can damage a child's health, even if they're 15 or older. It can affect the normal development of vital organs and functions, including the brain, liver, bones and hormones. (page 12)

- If 15-17-year-olds drink alcohol, it should be rarely and never more than once a week. They should always be supervised by a parent or carer. (page 12)

- Alcohol dependency can be called a disease because the person is at a dis-ease with themselves. Like any disease, it needs to be treated. Without professional help, an alcohol-dependent person will probably continue to drink and may even become worse over time. (page 15)

- One in five children in the UK lives with a parent who drinks too much – that's over 2.5 million children. (page 19)

- Almost no Local Authority is increasing its drug and substance abuse treatment budgets, despite the increases in alcohol-related hospital admissions. Of the 49 Local Authorities providing data on future treatment budgets, 70% (34 Local Authorities) are experiencing rising alcohol-related hospital admissions. (page 19)

- 19 per cent of 16–24s don't drink, and 66 per cent don't feel alcohol is important to their social lives. (page 21)

- People who add energy drinks to alcohol have a higher risk of injury from car accidents and fights, compared to those who drink alcohol straight. (page 22)

- Alcohol is created when grains, fruit, or vegetables are fermented. Fermentation is a process that uses yeast or bacteria to change the sugars in the food into alcohol. Fermentation is used to produce many necessary items – everything from cheese to medications. Alcohol has different forms and can be used as a cleaner, an antiseptic, or a sedative. (page 24)

- Girls or guys who have strong self-esteem are less likely to become problem drinkers than people with low self-esteem. (page 25)

- A study in the US has found that while 26% of young adults between the ages of 21 and 23 had seen an alcohol advertisement, 23% of 15- to 17-year-olds had also seen the same advert. They also found that young people who could accurately identify alcoholic products and who said they liked the ads were more likely to try drinking or to drink more. (page 39)

Glossary

Alcohol

The type of alcohol found in drinks, ethanol, is an organic compound. The ethanol in alcoholic beverages such as wine and beer is produced through the fermentation of plants containing carbohydrates. Ethanol can cause intoxication if drunk excessively.

Alcohol by Volume (ABV)

ABV is a measure of how much pure alcohol is present in a drink. It is represented as a percentage of the total volume of the drink. For example, a one-litre bottle of an alcoholic beverage will provide an ABV value on its label. This informs the buyer what percentage of that one litre consists of pure alcohol.

Alcohol dependency/alcoholism

Alcohol is a drug and it is addictive. If someone becomes dependent on drink to the extent that they feel they need it just to get through the day, they may be referred to as an alcoholic. In addition to the various health problems related to alcoholism, an alcoholic's relationships and career may also suffer due to their addiction. They can suffer withdrawal symptoms if they don't drink alcohol regularly and may need professional help from an organisation such as Alcoholics Anonymous to deal with their dependency.

Binge Drinking

When an individual consumes large quantities of alcohol in one session, usually with the intention of becoming drunk, this is popularly referred to as 'binge drinking'. It is widely accepted that drinking four or more drinks in a short space of time constitutes 'bingeing', and this can have severe negative effects on people's health.

Depressant

A substance that slows down the nervous system, making the user feel calmer and more relaxed. These drugs are also known as 'downers' and include alcohol, heroin and tranquillisers.

Hangover

A hangover describes the effects of alcohol the day after intoxication. Alcohol is a depressant, causes the body to dehydrate and also irritates the stomach, so hangovers usually involve a severe headache, nausea, diarrhoea, a depressive mood and tiredness. There are many myths about how to cure a hangover but the only real solution is to drink plenty of water and wait for it to pass – or of course to drink less alcohol in the first place!

Intoxication

The state of being drunk, caused by drinking too much alcohol. Drunkenness can lead to dizziness, sickness, loss of memory, aggression or anti social behaviour, as well as potentially causing long-term health problems such as cirrhosis of the liver. Due to the loss of inhibitions associated with heavy alcohol use, it can also cause people to indulge in risk-taking behaviour they would not normally consider – for example, having unprotected sex.

Teetotal

A teetotaller is someone who abstains completely from alcohol. If an individual is trying to recover from an alcohol dependency they will usually be teetotal. However, people do not drink for many other reasons, including religion, pregnancy, for health reasons or just through personal preference.

Unit of alcohol

The unit system is a method used to measure the strength of an alcoholic drink. One unit is 10ml of pure alcohol – the amount of alcohol the average adult can process within the space of one hour. Units can be calculated by multiplying the amount of alcohol in millilitres by the drink's ABV, and dividing by 1,000.

Assignments

Brainstorming

⇨ In small groups discuss what you know about alcohol.

- What types of alcohol are there?

- What problems can alcohol misuse cause?

- What are the health risks involved in drinking too much?

Research

⇨ Do some research into the types of cancer alcohol consumption might cause. Write a short report and produce a graph to show your findings.

⇨ In small groups, research alcohol guidelines in this country. You should consider the risks involved if people drink above the recommendations. Produce an infogram showing your results and share with the rest of the class.

⇨ Research alcohol abuse and the effects it can have on a person. You should consider their age and gender. What differences are there, if any, in the effects it might have on their health and behaviour. What are the risks involved in excessive drinking? Write a two-page report on your findings.

⇨ Talk to family and friends about their alcohol consumption and find out the reasons why they drink. Think of six questions to ask them and then write a report no longer than one page long.

⇨ In pairs, do some research into children who live with alcoholic parents. How does this impact on their lives? Is this likely to lead them to become drinkers? Share your findings with the rest of your class.

⇨ Do some research into the different types of energy drinks which are on the market. When you have gathered your results prepare an infogram.

Design

⇨ Prepare an advertisement for any brand of lager that will be displayed in a sports stadium.

⇨ Create a leaflet which raises awareness about the dangers surrounding excessive drinking. It should list some of the organisations which could be of help as well as practical advice.

⇨ Design a poster to be displayed in schools, which highlights the risks of drinking too much as a teenager. It should be informative and list the potential health problems and risky behaviours which can occur through excessive drinking and the long-term implications to their lives.

⇨ Design an illustration to highlight the key themes/messages of the article on page 12.

⇨ Think of a name for a new brand of lager which is due to come on to the market. You should design a label for the bottle.

Oral

⇨ Have a class discussion about drinking amongst teenagers and the risks to them. DIscuss ways in which you think these issues could be adressed.

⇨ In small groups, prepare a PowerPoint presentation that explains the health risks of drinking. Share your findings with your class.

⇨ Split the class into two groups. One group will argue in favour of alcohol advertising at sports events and the other group will argue against.

⇨ In pairs, go through this book and discuss the cartoons you come across. Think about what the artists were trying to portray with each illustration.

⇨ In pairs, stage a discussion between two friends where one of you is trying to persuade the other to give up drinking. Take it in turns to play the role of the persuader.

Reading/writing

⇨ Imagine you are an Agony Aunt writing for a national newspaper. A young boy has written to you as he is living with an alcoholic father. His father becomes argumentative and sometimes violent. Write a suitable reply giving advice and information on where he may look for support and help.

⇨ Write a one-paragraph definition of 'alcoholism' and compare it with a classmate's.

⇨ Write a blog about binge-drinking in the under-16's and explain why teenagers should not drink at such a young age.

⇨ Choose an article from the book and write a summary of it. This should be at least two paragraphs long.

⇨ Read the article on page 26 and write down your thoughts regarding this issue. Consider the following:

- Why do you think young people are shunning alcohol?

- Which gender is more likely to be teetotal?

⇨ "Cutting alcohol ads in sport sends the right message to youngsters". Write an essay exploring this statement. You should write at least two sides of A4.

Acknowledgements

The publisher is grateful for permission to reproduce the material in this book. While every care has been taken to trace and acknowledge copyright, the publisher tenders its apology for any accidental infringement or where copyright has proved untraceable. The publisher would be pleased to come to a suitable arrangement in any such case with the rightful owner.

Images

All images courtesy of iStock except pages 17, 18, 23, 29, 32: Morguefile, 20: Pixabay, cover image Jackie Staines

Icons

Icons on pages 8 were made by Pixabay.

Illustrations

Don Hatcher: pages 4 & 25. Simon Kneebone: pages 14 & 38. Angelo Madrid: pages 6 & 34.

Additional acknowledgements

With thanks to the Independence team: Shelley Baldry, Sandra Dennis, Jackie Staines and Jan Sunderland.

Tina Brand

Cambridge, January 2018